Bob Crowder

Bob Crowder

A New Zealand organics pioneer

MATT MORRIS

OTAGO UNIVERSITY PRESS
Te Whare Tā o Te Wānanga o Ōtākou

Published by Otago University Press
Te Whare Tā o Te Wānanga o Ōtākou
533 Castle Street
Dunedin, New Zealand
university.press@otago.ac.nz
www.oup.nz

First published 2024
Copyright © Matt Morris
The moral rights of the author have been asserted.

ISBN 978-1-99-004874-6

A catalogue record for this book is available from the National Library of New Zealand. This book is copyright. Except for the purpose of fair review, no part may be stored or transmitted in any form or by any means, electronic or mechanical, including recording or storage in any information retrieval system, without permission in writing from the publishers. No reproduction may be made, whether by photocopying or by any other means, unless a licence has been obtained from the publisher.

Editor: Imogen Coxhead
Indexer: Diane Lowther

Front cover: Main photograph: Bob Crowder Collection; Artichoke illustration: Shutterstock

Printed in China through Asia Pacific Offset.

Contents

7 Preface

15 1: Britain, 1939–62

33 2: Auckland and Pukekohe, 1962–66

49 3: Lincoln and the Horticultural Research Area, 1966–76

71 4: Organic conversion, 1975–83

95 5: Developing organics, 1984–87

117 6: New Zealand organics on the world stage, 1988–92

133 7: IFOAM '94

147 8: The struggle to save the BHU, 1995–98

161 9: The death of the BHU? 1998–2000

179 10: Retirement, 2001–23

200 Notes

231 Bibliography

235 Index

Preface

I first met Bob Crowder around 1993. Our initial conversation took place in aisle seven of the New World supermarket in St Martins: toilet paper and pet food. I was 18 or so and worked at the supermarket part time while studying history at the University of Canterbury. Bob was wearing an open-necked shirt and shorts and carried a shopping basket with just a few things in it. I often saw him there, and he never shopped with a trolley. It didn't occur to me at the time, but of course he only used the supermarket to supplement the abundance of food pouring out of the Biological Husbandry Unit (BHU) at Lincoln: he would have done his main shopping at Piko Wholefoods. His main purpose in coming to New World was to deliver vegetables to the produce department.

It can be quite boring working in a supermarket, and customers rarely take an interest in the staff. Bob was different. He engaged me in a lively conversation. When he learned that I was interested in Roman history, right there beside the dog food he explained that Rome had collapsed due to the over-exploitation and desertification of the environment. He left me with his card and told me with wide eyes that he was on the world board of IFOAM. At the time this meant nothing to me. I later found out it was the International Federation of Organic Agricultural Movements, an organisation that had been driving and supporting the development of organic production across the globe for more than two decades.

I planned to travel to Italy, Egypt and Britain in 1995. Bob was intrigued by this and impressed, I think, that I was using my supermarket labour to fund the trip. I had no idea that Bob, at 55, was engaged in the biggest year of his life – the organisation and delivery of the IFOAM '94 conference.

I got to know Bob properly in 1995. Visits to his enormous, vibrant and abundant garden were precious windows in a fairly bleak period of my life. We ate globe artichokes. It was the first time I'd eaten them: Bob had to teach me what to do. I don't think he mentioned that the artichokes were his own 'Crowder's Delight'.

Bob's conversation was stimulating and affirming: I felt he actually saw who I was. This rare gift appeared to come naturally to him.

To me, the world seemed quite dull, but Bob lived in a different reality. I have now read decades of his numerous diaries. Invariably, he comments on the weather: a bad day is characterised as 'anticyclonic gloom'; the best involves forked lightning (rare in Christchurch). A monocultural paddock, tree rings sprayed out with herbicide, leaves him cold and despondent; an orchard filled with flowers and insects makes his heart sing. He wasn't interested in tacky, meat-market-style gay bars, but an afternoon of Morris dancing or a night in a jazz bar would fill him with joy. Bob did not do boring.

I write this because there is an obvious bias present in this book. After the artichokes in 1995, in 1997 I moved into Bob's sleepout. I met Brendan Hoare and began to meet others in the organic and Green movement. In 1999, having completed my honours year, I wanted some time out before starting a PhD. Bob encouraged me to apply for the coordinator's job at the Organic Garden City Trust and was delighted when I was offered the job. From there I quickly became involved in the Soil & Health Association (of which I became national chair), and before long I was coordinating the local pilot of a publicly funded organic certification scheme now known as Organic Farm NZ. In 2009 I started working at the University of Canterbury in the Sustainability Office. I have been immersed in this world ever since and was a first-hand witness to some of the events described here. Bob's guiding hand is clearly visible in all of this, and I write this book out of gratitude to him.

Bob was a lecturer in horticulture at Lincoln University, and he was also tasked with developing Lincoln's original Horticultural Research Area in the 1960s. He went further and in time transformed it into the only university-based organics research unit in the country: the Biological Husbandry Unit. He has certainly been a divisive figure at times, and his fiery personality and loathing of bureaucracy made university life challenging for him. Diplomacy too often looked like hypocrisy to Bob, and it is clear from my own experiences, from the interviews I have conducted for this book, and from Bob's diaries and many other sources, that Bob says exactly what he thinks. At times this can be stinging and hurtful, and sometimes he has lost personal and professional relationships as a result. On the other hand, he has an endless capacity to try again, even if others don't. When hoped-for reconciliations have not transpired, he has appeared confused. In meteorology, a lightning storm clears the air; not always so with human relationships, it would seem.

But while there have been lightning flashes along the way, as there always are in life, Bob's formidable achievements have inspired countless people to succeed. And he has always delighted in that.

Many of those people are present in this book. I want to thank everyone who participated in interviews and answered follow-up questions. Although this book is focused on Bob, it is also framed around a critical period in the development of the commercial organic industry, both nationally and internationally. That story has not yet been told fully, and I know some of my interviewees would have liked me to explore more of this in my study. They have provided tremendous insights that will perhaps be picked up at a later date.

Some readers may notice that the majority of interviewees are male. There has certainly been comment from some quarters that Bob did not like or associate with women. This is not true, but perhaps women needed to work harder to get his attention. It would be wrong to say that Bob segregated himself from women: indeed, women figure prominently in his closest circle of friends, and female volunteers for a long time were critical to the upkeep of the BHU. I suspect, though, that Bob felt more comfortable in male company, where perhaps there is less expectation to discuss matters of the heart, or family, or other delicate subjects. If he were to talk about these, it would probably be with a select number of trusted gay men.

I would also like to thank those historians who have read this manuscript with a critical eye. This book builds on a modest but growing literature on the history of organic farming, which includes William Lockeretz's *Organic Farming: An international history* (2007) and more recently Gregory Barton's *Global History of Organic Farming* (2018). This wonderful contribution to the study of farming around the world elucidated the significance of many early players in the organic movement, particularly the Howards, but shone little light on how the movement played out in New Zealand, or how New Zealand's experience impacted the rest of the world. Although New Zealand is a small country, this was a notable omission, if only because it was in New Zealand that Sir Albert Howard's work was first adopted by a national organisation: the New Zealand Humic Compost Club.[1]

There still awaits a proper history of the organic movement in New Zealand. There have been several preliminary attempts, among them Claire Williams' master's thesis, 'The New Zealand Humic Compost Society' (1985); interview excerpts in Scott McVarish's *The Greening of New Zealand* (1992); passing references in both editions of Pawson and Brookings' *Environmental Histories of New Zealand*; my PhD thesis 'A history of Christchurch home gardening from colonisation to the Queen's visit: Gardening culture in a particular society and environment' (2006); Bee Dawson's *A History of Gardening in New Zealand* (2012); Guilherme Barreto's master's thesis, 'A new beginning: The establishment of the biodynamic movement in New Zealand 1930–50' (2018); and most recently my own *Common Ground: Garden histories of Aotearoa* (2020).

The last of these tells some of the story of the New Zealand Humic Compost Club and its later incarnations as the New Zealand Humic Compost Society, the New Zealand Organic Compost Society, the New Zealand Soil Association and the Soil & Health Association of New Zealand. Founded in 1941, it was a bastion of those who believed in the links between soil health and human health. It had a heroic early history in the Dig for Victory campaign of World War II, but by the 1970s had become a series of garden clubs supported by a fairly stuffy bureaucracy. When Bob encountered the organisation, he was not inspired, but it soon became an enormous support to him. The rich story of the Soil & Health Association is yet to be fully told.

This book is about more than Bob's contribution to horticulture, however. Bob grew up in Britain during World War II and was a schoolboy in the 1950s at the time of the gay scandal of Lord Montagu.[2] Bob's journey of coming out to himself in the 1960s is long and contorted – as it was for many gay men of this period. A wonderful assortment of diaries and letters helps to unpack this early part of his life and offers rich insight into the world of a man wrestling with his sexuality – and the impact this had on his public life. Not until the 1970s was Bob able to fully accept this aspect of himself. When he did, he flourished.

In another sense, though, the story of Bob's sexuality is very much connected to his contribution to horticulture. His story helps to shine light on the New Zealand gay community in the 1960s and 1970s and the crushing heterosexism and homophobia of this period. And few places conform more to the straight, white, male paradigm than an agricultural college, which is of course where Bob landed. There is apparently almost no literature on the 'queering' of agricultural – let alone horticultural – science, although a small body of work identifies the ways in which some queer farmers have carved out a safe niche in the world of 'sustainable agriculture' (which includes organic farming). Even then some have argued that equating sustainable farming with family farming leaves queer farmers still scrabbling to conform to heterosexual mores.[3] I contend that Bob's discovery of organic farming in the period from 1977 was one means of empowering himself by reconnecting to his core values, but Lincoln University's culture of heterosexism and homophobia seized on organics in covert ways in order to marginalise both Bob and the science behind organic horticulture. That is to say, heterosexism and homophobia acted as constraining forces that impeded science. Not that Bob would agree: he viewed Lincoln as a liberal institution. At the same time, however, he was never invited to the kind of work dinners that male staff took their wives to and where opportunities for advancement were subtly rehearsed.

Aside from any raised eyebrows about Bob's personal life, there was also antagonism regarding the notion of organics, which many considered contradictory to axioms of science and economics. The economic imperative behind New Zealand's farming strategy was widely regarded as common sense, and in challenging this Bob made himself a target. At best he was marginalised as naïve

Bob and sunflowers in his garden.

or eccentric, at worst as a renegade or even a 'wrecker'. There is much debate about the vested interests of multinational agribusiness and its ruthlessness in destroying the livelihoods of small-scale organic farmers. That debate is beyond the scope of this book. However, Bob did observe (and experience) derision from that quarter too, but nevertheless continued to press his case.

Bob was, and is, an immensely social person, but has not been without his complexities and contradictions. He has often expressed the thought that he is 'in the world but not of the world', 'outside the dominant paradigm', 'an outsider'. He has had a few partners but sometimes struggled with commitment. Professionally, he felt misunderstood and derided by his so-called colleagues. He was a caring and dutiful son but did not relate with members of his family at a deep level – except perhaps his Uncle Cyril (whom he never met), who also left England for New Zealand and was known as the black sheep. Politically, he has fought for the underdog and lived a fairly simple life as an ardent socialist.

And then there was his relationship with God and the Church. Although this was strong when he was a young man, it later melted into feelings, experiences and expressions that didn't need codifying, such as seeing the world from the top of a mountain or Morris dancing among spring daffodils. As he conveyed to Scott McVarish in 1992:

> Spiritualism is appreciating the meaning of life and the whole feel of the place where you live … I introduced Morris dancing from Britain to share a bit of spirituality and the meaning of life to Christchurch. It is people making their own entertainment, enjoying being together and all revolving around the natural aspects of fecundity, fertility and the sun.[4]

His spiritual vision of a simple life enriched by an enlivened environment stands in stark contrast to his view of how most Westerners live:

> Western lifestyles are based on rampant exploitation of the rest of the world … [but] people in the Third World are starving because of our demands on the global environment. If we really want to help, we have to cut our demands.[5]

The Christian Church ultimately did not feel like his natural home, but the message of doing what we can to alleviate suffering and offer care to the weakest among us, in order to build a kind of new Jerusalem, is a good fit.

In all of this can be seen the journey of a man looking for his proper home. Even later in life Bob embodied certain ideas that might be considered contradictory by some. He was a fan of Prince Charles (now King Charles III) but also a socialist. He helped establish an export market mechanism for the primary industries but maintained a longing for the parochial agrarian life crystalised in the Morris dancing movement. The expression of his core beliefs changed significantly over his life, but never the core beliefs themselves. He gradually distanced himself from certain institutions with which he had been heavily involved – like the Church, like the agrichemical industry – when it became clear they could not align with his developing views. He experimented with new 'homes' for his beliefs, such as the anti-Vietnam War movement, the gay world and the folk scene. But ultimately, I think, no pre-existing structure or community was complete enough to support him.

That is, until he encountered the nascent international organic farming movement. In New Zealand this was so underdeveloped that there was nothing else for it: it was a matter of personal survival for Bob to ensure the organic world could come to life around him.

For this effort we can be grateful. Since the late 1970s, when Bob began experimenting with organics, the argument for sustainable farming has become stronger than ever. Unsustainable farming (including systems reliant on synthetic pesticides, fertilisers and petrochemicals, the drivers of unprecedented deforestation and overstocking of livestock) has caused incalculable damage to our planet and our wellbeing. Bob's foresight, then, should be acknowledged as both wise and prophetic.

I thank the wonderful team at Otago University Press for making this book possible, and Bob for his patience with me as I went through the long-winded process of piecing his life together and for giving me unlimited access to a vast array of materials to work with. Bob has been assiduous in keeping documents associated with all aspects of his life, and the enormous cache of diaries, letters, faxes, meeting minutes and other items at my disposal was both daunting and refreshing. The final product cannot do justice to the details contained in these papers; many important moments and people could not be captured at all.

Bob takes great interest in people all over the world. He has always been someone who cared. It has taken me years to write this book; I hope that the final product reflects this care back to him and does some justice to his important and influential life.

TOP LEFT: *Madge and Bill on holiday in Bournmouth, 1930.*
ABOVE: *Bob, David and Peter the dog, Medina Avenue, Shide, c.1946.*

1: Britain, 1939–62

On 4 January 1939 the weather was dreadful. Most of the United Kingdom was inundated with rain, and snow had been falling heavily in many places since New Year's Day. For England and Wales, this was the wettest January 'since before 1868 and as far as can be estimated since 1764', or so the weather report said.[1] In the Birmingham area, an intense snowstorm was raging. In the midst of this tempest, in Solihull Hospital, a new building constructed on the site of the old town workhouse for 'the undeserving poor', thirty-year-old Madge Crowder gave birth to her second child, Robert Anthony Crowder – the Bob of our story.[2]

Robert's father, Bill, was a builder. When Robert was barely six months old, Bill got a job in Newport on the Isle of Wight. For Madge and Bill, this was a big move. The Crowder family had lived in the Birmingham area for more than a century. It was hard for a young couple with two small children in tow to leave their parents behind. And then on 1 September 1939, Germany invaded Poland, and World War II commenced. The National Services (Armed Forces) Act was passed promptly: all men aged 18 to 41 were liable for conscription. Bill was 31; once conscripted, he trained as a radio operator near Bath. Madge remained in Medina Avenue in Shide on the Isle of Wight with toddler David and baby Robert.

The Isle of Wight endured numerous German bombing raids. Bombing and 'strafing' – where Germans flew down the railway corridors shooting at civilians – were common occurrences, and Robert was frequently terrified. When the bombs fell, Madge and the children would huddle in a mattress-lined alcove under a bookcase in the living room with Mrs Davis, a woman who lived across the street. In 1943 Medina Avenue was bombed. The row of houses over the road, including the Davis house, were destroyed.[3] Madge and Bill's own home was significantly damaged. Bill came home briefly from Bath to help sort out living arrangements. For four-year-old Robert, the effect of the blast was immense. He went into shock, began losing weight and was soon in hospital, where he remained for several weeks.[4] Little Robert felt very alone in his high-sided cot and later recalled that no one ever seemed to be around when he needed them.

Meanwhile, his six-year-old brother David perfected the art of listening for the V1 ('Doodlebug') rockets, which the Germans used from the summer of 1944. By then households had been issued with indoor air-raid shelters and outdoor air-raid shelters that were shared between two or more homes. The Crowders had access to a shelter in the Benningtons' garden next door through a gate in the fence. One evening the air raid siren went off. Madge had begun to dress the children to go to the outdoor shelter when David suddenly raised a finger and said, 'I can hear them!' Moments later the *chug chug chug* of rockets became obvious. Madge bundled the boys up and raced outside. They were at the gate when David cried 'They've stopped!' just before the bomb struck and they were blown into the shelter.[5] Everyone survived but it was another terrifying experience that Robert would never fully shake off.

This event precipitated a temporary return to Solihull, where Madge's father had died earlier in the year and Bill's father was on his deathbed. The journey was arduous. Madge, with her two sons and Peter the dog, needed to cross the Solent to Portsmouth and from there catch trains to Solihull. As the family waited at one bombed-out platform, a train full of American GIs passed. Robert was astonished when the men opened the windows and showered them with sweets.[6] At Solihull Robert met his paternal grandfather, Albert Crowder, who was bedridden and died not long afterwards, in December 1944. Robert considered his grandmother Elizabeth a 'Victorian grandmother'; she was 'deeply religious' (this was not a compliment) and she vigorously shut down the boys' attempts at fun and games.[7]

While in Solihull they must have visited Madge's mother Emily, now living alone. It is likely that this is when a decision was made that Emily would come to stay in Shide. By this time, the war effort had turned, and, following a wave of Allied victories in Europe, Emily felt safe enough to travel from Solihull to Portsmouth Harbour, where Madge met her and accompanied her to Shide. Emily had already sent her luggage ahead, which suggests she was not making a short visit.[8] She was now a fervent Jehovah's Witness and had vigorous arguments with Madge about religion. Emily brought a certain level of violence into the house and would chase the boys with her stick when she thought they were up to no good. Robert did not appreciate her controlling ways.[9] Emily died not long after and left all her money to the Jehovah's Witnesses, which was a huge setback for the young family and imbued Madge with deep cynicism about religion. She henceforth refused to enter a church; a move that foreshadowed the kind of churchless spiritualism Robert later evinced.[10]

Most of the beaches on the Isle of Wight were closed during the war, apart from Gurnard, to which they sometimes travelled by bus.[11] Huge metal pipes had been stuck into the sand to repel invading landing craft and made a great playground for small people. On one of these trips, from the top of the double-decker bus, Robert saw a dogfight in the sky and watched a German fighter plane being downed.[12]

In 1946 Bill came home at last. He had been involved in the Burma campaign, flying supplies to the Chindits – a British–Indian combination that penetrated beyond the Japanese lines. The Chindit operations ceased in 1944, but Bill had remained in India on other missions. Robert, now six, did not know his father at all and was somewhat anxious around this stranger who seemed to want to take his mother away. He sought fatherly connections with other men in his life, including his teachers. Having Mr Hayward's arm around him was calming, reassuring. Likewise, Mr Kenny's supportive hand under the small of Robert's back in the freezing swimming pool was a welcome and comforting feeling.

Mr Kenny also took the class on memorable walking trips. One of these was to Niton at the southernmost tip of the Isle of Wight, to see St Catherine's Lighthouse. Originally built in 1323, it is one of the oldest lighthouses in Britain. The class were swimming among the rocks at the bottom of the cliff when much to Robert's surprise his parents suddenly appeared. They had been alerted to gossip about Mr Kenny's purported predilection for small children and had come to keep an eye on things.

Gradually, a sense of normality began to return. Mrs Davis's property across the road remained a site of desolation for years – even in 1961 most of the section was still vacant.[13] Robert, his brother and 'the gang' hung out there and considered it a great playground.[14] Although Mrs Davis's house had been destroyed, the string board of her grand piano was still partly intact and the children enjoyed dancing on the strings. Sections of the cellar were also still extant, complete with an adder that had fallen in, which the boys felt compelled to stone to death.

In 1947 Britain experienced some intense extremes of weather. The year opened with unusually cold temperatures, reminiscent of the year of Robert's birth. The cold intensified in February – this was now the coldest winter since the late nineteenth century.[15] Heavy snowfalls shut down many areas of England, and sub-zero temperatures persisted for long stretches of time. On top of rationing, railway strikes and a fuel crisis, and with many communities still in a state of disrepair following the war, conditions were miserable.[16] The weather was so extreme that the Crowders' macrocarpa hedge died.[17] Robert was confused by this, because the weather reports he was reading said the snow was coming on an anticyclone from Siberia, meaning from the south. This collided with the understanding he'd had from the children's song that declared 'The north wind doth blow, and we shall have snow, and what will poor Robin do then? Poor thing.' He looked into this contradiction with fascination; the experience left a lifelong impression.

With the beginning of spring in March, snow melt and high rainfall caused 'unprecedented' flooding across much of England and Wales, which put paid to any idea of bountiful harvests later on.[18] An extremely hot summer followed with numerous thunderstorms through June and parts of the highly unsettled July. August 1947 was the hottest on record, and the dry weather continued into autumn with unseasonably high temperatures, even in November. Now there were drought conditions.[19] Robert's already considerable curiosity about things meteorological was piqued.

Robert's parents spent time in their rather wild garden, which featured four or five old apple trees and many large trees and shrubs that provided a lot of shade and possibly made vegetable gardening difficult. The greenhouse windows were shattered by bomb blasts and never repaired. Robert started a garden of his own on Mrs Davis's property, close to the remains of the piano, on the site of her old garden. He and his father pored over seed catalogues together and Bill placed the orders. Robert began with sweet William (which Madge loved), gaillardias and lupins. In 1947, when he was eight, he began growing vegetables in earnest.[20] The combination of food rationing and poor harvests may have been his inspiration: as he said, 'I got fed up with swede.'[21] Nutritious powders sent to the family by their American relatives alleviated things only so much.

There is a significant literature around the deliberate use of gardening as a healing practice: horticultural therapy. There is also plenty of evidence that making a garden can be an intuitive response to trauma, perhaps especially when the land itself has been traumatised through either natural disaster or warfare. This point is well made in Lalange Snow's *War Gardens* (2018).[22] Snow quotes an Israeli woman who was working in a garden therapy programme in Sderot, an Israeli town close to Gaza known as 'the bomb shelter capital of the world': 'Gardening is one of the most effective ways of re-earthing yourself, being calm and not a fanatic.'[23] After visiting many gardens in some of the most war-torn parts of the world, Snow reflected that 'to garden in a time of crisis and conflict is to escape to "a world within a lost world" … It is a refusal to accept a world defined by violence and destruction but instead create life.'[24] Eight-year-old Robert no doubt found the work therapeutic, and his garden crystallised a sentiment that would drive him for the rest of his life, as he reflected in 2018: 'I can see the runner beans even now hanging on their sticks in this derelict bombsite … All those things had an extremely strong influence on me in directing my life.'[25]

The Davis property had a lot of rubble to contend with but good alluvial soil, and Robert had the support of his parents and inspiration from the Gullivers next door, who had a big orchard. Encouragement also came from Mr Joliffe, who lived on the other side of the Davis property. Joliffe had 'a nice country garden … He was a great vegetable grower.' His property bordered the river, and as the hot summer and dry autumn dragged on, access through his property to the river was a boon. Robert lugged many buckets of water up the hill and experienced back pains as a result of this exertion.[26] His garden was a huge success: he supplied his household with vegetables and kept this up until they left the Isle of Wight in 1949.[27]

Robert also learned from Mrs Belben, a deaf and mostly mute woman who lived on Burnt House Lane, on the way to an old chalk pit where Robert liked to walk his dog.[28] Her house was set among massive trees that kept her wild garden sanctuary in delicious shade. Somehow Mrs Belben and Robert struck up a friendship, and soon he was helping her create a vegetable garden in an area where there was enough light. Mrs Belben gardened with him and he learned from her.[29] There is something poetic about a boy being taught to garden by someone who cannot speak – at least not in words. Gardening, as he would always know, must be practised, not preached.

Perhaps the peaceful pursuit of gardening provided a necessary counterpoint to the rougher times Robert and his brother had with the local gang. Colourful characters among this group included a 'tom boy' – a girl who lived in a shop near the Crowders and entertained the boys by exposing herself under the bridge. She introduced them to stolen cigarettes; Robert, like the others, turned green and was violently ill (from the smoke, not the exposure).

Pitched battles raged in the bombsite across the road between rival groups who set up barricaded forts, hurled stones or deflected them with shields made from sheets of iron. It was 'pretty brutal'. In one battle David was pushed into a large patch of nettles; the parents in the street raced around collecting dock leaves to smear him with. The best of all the play locations was the chalk pit itself. The disused rail infrastructure was still intact with railway tracks, functioning points and a trolley, which they lifted onto the tracks and drove back and forth through the tunnel. It was great fun, dangerous and wild.

The Crowders were good friends with the Chivertons, who also lived on Medina Avenue, and after the war the families buddied up for foraging expeditions, including for blackberries. Everyone carried sticks to rattle in the brambles to frighten away the adders before stepping in. Adders were plentiful: the scrub that had grown up during the war years had enabled them to proliferate.

The families made many trips to the beach together by train. The children enjoyed leaning out of the windows and getting soot in their eyes. On one occasion, they narrowly avoided falling out of a door that opened suddenly: Madge grabbed one of the girls' dresses and yanked her back in the nick of time. Then would come the walk from the station to the beach – the Chivertons with a pram, the adults carrying the picnic, swimming gear and other paraphernalia. Down the

CLOCKWISE FROM TOP LEFT:
Bob with Peter the dog, Sandown, 1947; Bob at school in Newport, Isle of Wight, c.1949; Bob in his garden, Avon Road, Devizes, c.1951; David and Bob, probably Bournmouth, c.1950.

promenade at Sandown they went towards Culver Cliff, where a popular canoe lake had opened. The children played in the boats while the parents enjoyed tea and cake in the tearooms.

The Crowders spent Christmas holidays back in Solihull at the family home with Elizabeth, now in her seventies. Christmas time was when Bill and his brothers got together, and it was a chance for Robert to get to know his uncles. David and Robert found Elizabeth quite challenging. She was a strict Anglican and on Sundays they weren't allowed to do *anything*. If Uncle Bert had the television on, she would sit with her back to it.[30]

Robert began to attend secondary school in Newport around 1949. Among his memories of the brief experience was having to tend a garden plot there, an unsatisfying experience given the shady site and sticky, wet soil. Nothing thrived. His Newport school memories are few because later that year Bill got a job in Devizes, Wiltshire. The family prepared to move. It was hard for David, who wanted to live near the water so he could pursue his interest in sailing, but 10-year-old Robert was keen to go.

Devizes had a population of just under 8000.[31] It was an old market town and even in the 1950s was known for its snuff production and brewing.[32] By then, however, it was transforming into a centre of manufacturing. The opportunity for Bill came in this changing context: his job was to help set up a kitchen and bathroom company in which he became a specialist in top-of-the-line AGA cookers.

Robert initially attended primary school but following the 11+ examination he gained entry into Devizes Grammar School.[33] He thrived. The boy who loved to garden and was fascinated by weather became a rugby player (obliged to play, against his wishes, due to his physical prowess), a cross-country runner, a tennis player and a dancer.[34] He took the lead in several drama productions, skated on the Kennet and Avon Canal and went fishing often. He enjoyed a healthy social life but, content with his own company, did not form many close friendships.

Bill's work installing AGAs took him out into the farmlands around Wiltshire and in the holidays Robert would join him. He loved exploring the area and seeing how country folk lived. Robert grew to know the Wiltshire countryside well. On long walks with his dog, he observed the natural world around him. He always knew where the first spring flowers were – Pottern Woods for the first primroses and bluebells – and brought back posies for Madge.

At other times he was immersed in natural history books – or Biggles, which he also enjoyed. He became a voracious reader, and when he ran out of books to read at the local library, his parents enrolled him with the Boot's library service.

As hormones kicked in, he occasionally experienced fleeting feelings for some of the boys, but his attention was taken up with gardens, sports and the weather. In addition, any such feelings towards other boys were crushed during the Lord Edward Montagu trial of 1954, when Robert was 15. Montagu had been arrested on two charges of having underage sex with a boy scout in 1953 (which he always denied) and again in 1954 for indecent behaviour with two RAF servicemen, for which he was imprisoned for 12 months. It was all very 'hush hush', though everyone at Devizes Grammar seemed to know about it.[35] Gay filmmaker Derek Jarman, who was born three years after Robert, recalled that throughout the early 1950s 'a moral crusade fuelled by the hostile attitude to homosexuals in the McCarthy show trials led to a spate of arrests' that culminated in the Montagu trial.[36] 'I was dimly aware,' Jarman wrote, 'that this national scandal related to me, though I had no words to describe it.'[37] It may have been similar for Robert.[38] Certainly it was the case for one of Robert's schoolmates, David, who had fallen in love with their mutual friend Graham, and who later recalled, in a letter to Bob:

> At 16 I was having two years of psychiatric 'help' to cure me of being gay. At the time the newspapers were full of hate towards gay people and I was almost suicidal. The book by Peter Wildeblood tells the story very well, as it was he and Lord Montagu who in some ways precipitated the national orgy of hate following the incident with boys in your year ... on a scouting trip.[39]

In later life Robert described his sexual life as 'sublimated'. Deploying the Freudian use of this term in this way helped him to make sense of both his sexuality and his social life. Sigmund Freud claimed that humans have a capacity to displace sexual energy and put it at the disposal of 'civilised society' through a process of sublimation.[40] This sublimation of sexual desire, in Freud's view at least, provided a necessary generative force within human society. As Jonathan Dollimore wrote, 'Civilisation actually depends upon what is usually thought to be incompatible with it.'[41] According to Freud, few people could master abstinence through sublimation; homosexuals, however, were 'often distinguished by their sexual instincts possessing a special aptitude for cultural sublimation'.[42]

Whether or not there was a process of Freudian sublimation at play, it is true that Robert buried his sexual feelings towards boys and instead focused intensely on the world around him: orchids in the chalk downlands, sport, the process of atmospheric pressures and, of course, gardening. All kept him more than occupied.

Robert displayed leadership skills, too. He formed a cross-country running team and started a school tennis group that met at the court close to his home – partly so he didn't have to play cricket. With the support of the woodwork teacher, he took his interest in weather to a new level by building a Stevenson screen (a shelter for meteorological instruments) as a class project, and initiated a meteorology club at school.[43] Robert had been watching the weather for several years now and had a good understanding of its mechanics. Geography, which covered all the things he was most interested in, became his best subject.

At the start of 1956 he began journalling. Rather than filling his diary with hormonal angst, Robert focused strictly on the weather. Each day he cut out the weather map from the newspaper, stuck it in his diary and added his own observations. The entry for 1 January 1956, three days before his seventeenth birthday, is typically detailed:

> The weather has remained mild and damp with considerable high cloud. Only a little very slight rain has fallen and a pale sun broke through at times.
>
> The weather is dominated by a rather complex depression with several centres which has brought only a weak SW … over us. A vigorous cold front is moving south and by evening was situated only a short distance to the north, it is moving only very slowly south.
>
> Behind it temperatures are 15°F cooler than at present and brought heavy snow to northern districts, 6" in Yorkshire.
>
> It is doubtful whether snow will fall here but there is quite a good chance of this cold weather spreading in tonight.[44]

Nor did he record anything of a personal nature on his seventeenth birthday: 'Pressure was dropping steadily this morning and a thick uniform layer of stratus or Nimbo Stratus covered the sky with considerable mist.'[45] He displayed slight excitement on 20 January at the prospect of ice-skating, and on the 23rd he wrote, 'I had my first skating this evening and so no more snow for me.'[46] On 29 January Robert noted, romantically, 'This evening was calm, clear and beautiful with a sunset which lit up the cirrus blooms in a deep greenish sky.'[47]

It wasn't long before the enterprising teenager began to contribute his observations to the local newspaper. The first of his monthly reports appeared in April 1956, and he kept them up until he left Devizes Grammar in 1958. In his initial column he noted that he had been keeping weather records since 1950 (he didn't say that he was 11 at the time).[48] Robert took his work seriously and wrote in a style that mirrored the met columns that he read so assiduously. His first published words reflect this:

> The month started off with dull mild weather with a good deal of hill fog and drizzle but soon it became dry and cooler with many days of sun. There was frost on ten nights with a lowest temperature of 25 degrees F. on the 13th, the highest minimum since 1952 … The average cloud cover was 53 per cent, four per cent less than 1955.[49]

It is no surprise that this young man was considering a career in meteorology!

Gardening remained his other passion. The family lived in a rented house on Avon Road for four years, and Robert developed an extensive garden there that gave him much satisfaction. He established a lawn and grew a range of vegetables, sourced mainly from the Sutton's Seed catalogues. The property bordered the railway line; on the other side of the tracks there were gardening allotments. Robert's neighbour, Mr Fishlock, had one of these. Robert observed Fishlock's extensive garden and in time acquired an allotment of his own on top of the embankment. Once again, he found himself carrying water up a hill to irrigate his vegetables (this time from a barrel that collected water from the family's roof). He set up his own rudimentary weather station at home – a maximum–minimum thermometer in a wooden box screwed to a post – and started taking daily readings.[50]

Robert's green fingers influenced his parents' next choice of a place to live. In 1954 Bill and Madge had a house built on a large section in a new subdivision on Queens Road. Fifteen-year-old Robert set to work. He built a terrace and dry-stone walls from Cotswold stone, planted a magnolia, espaliered apple trees and grew currants, berries and beans – always beans. The soil was good: upper greensand overlying chalk, light and very fertile. Of course, the met station came too. Bill and Madge were happy there and didn't move again until they emigrated to New Zealand in 1990.

The family maintained a strong connection with the Isle of Wight throughout the 1950s and for five years they visited every summer, renting a caravan at Whitecliff Bay. They were joined by the Chivertons, and the families enjoyed holidays together, exploring the sandy beaches and rocky headlands and dancing at the country club in the evenings, which Robert and David loved.[51] These family holidays were no doubt something of a consolation for David, who had been reluctant to leave the isle because of his passion for sailing. On one of these holidays, he sailed from Whitecliff Bay to Sandowne with Wilf Box, who had been Bill's pilot during the war. David went on to become an accomplished sailor.[52]

David was gregarious and had a strong group of schoolmates who shared an interest in cars. Robert was quite different. One of the few friends he made at school was Graham Hancock. The pair would go walking together looking for orchids, and Graham joined Robert's meteorology club and the cross-country running and tennis groups. Robert didn't feel any physical attraction to Graham, but he certainly felt this for some of the other boys, although 'nothing ever happened. I didn't allow myself to have close friends, and that was probably because I knew I was attracted to them and knew it was abnormal.'[53]

But there were crushes. One of these was with Mr Fishlock's grandson, who kept his hair beautifully slicked with Brylcreem and introduced Robert to the old-time dancing club in town. Robert went on to join the Patricia Scott School of Ballroom Dancing, where he met his dancing partner, Jean Clack. The pair won medals in a number of competitions.[54]

Robert enjoyed school, especially once he entered the academic stream.[55] His work was patchy, however. In his final year he failed French and one of his sciences but displayed a special aptitude for geography and natural sciences. He took geography as a scholarship subject on top of his A levels and won a distinction. It was enough to get him into university.

Still unsure of the direction he should take – meteorology or horticulture – he received rather direct advice from the maths teacher at careers day: 'Crowder, I think you'd be better off doing horticulture.' Robert took this advice and began applying for universities. He was interviewed at Wye and Nottingham and was accepted at both institutions. He chose Nottingham because it was hardest to get to from Devizes (and therefore created space from his parents, especially as his sexual identity developed), and after some extra tuition in French and passing a scholarship exam to pay for his place, his immediate future was set.

In May 1958 Robert contributed the last of his monthly weather columns to the *Wiltshire Gazette*, in which he wrote of the unusually cold month that had finished by delivering 'the hottest April weather for six years'.[56]

In July Robert and Graham Hancock set off on a week-long hitchhiking adventure around Devon. The trip included some highs, but also lows of driving rain and poor visibility, 'oozing and blistered feet' and 'nightmarish never to be forgotten hours', as Graham wrote. Apart from the ponies, they were 'the only living creatures' as they walked across Dartmoor, and by the time they got to Bath on the return leg, Graham confessed they were 'getting a little tired of each other's company'.[57] Robert, who thrived in the outdoors, whatever the weather, was not fazed. In letters to his parents he wrote, 'Having a good time despite the weather', and 'Had a swim in a sheltered bay where it was warm, its sea is icy cold but very invigorating.'[58]

Once at Nottingham, Robert found digs within walking distance of the university. He had to spend his first year at the main campus because he had failed a science paper and was required to take a preliminary course with physics, chemistry and biology. He found it quite easy. Perhaps too easy. With a light course load, Robert threw himself into sport and social life. During an orientation lecture he was advised to take up a sport to stay fit (and to stay out of the coffee bar as it would soak up money and study time), and in 1959 he joined the rowing club. It was the best sport he'd ever tried, and he excelled at it.

The summer of 1959 was a high point for Robert. He had new friends, such as Martin Taylor and Kenyan Geoff Nightingale. He recalled glorious weather and rowing and lazing around at the Lido. Back pain that he had first experienced as a child when carting water to his garden in Shide returned, but he rowed through it.

As a member of the University Biological Society, and because of his interest in meteorology, he was selected to take part in a trip to Tende in the southeast of France. The group was split in two: some went in a truck with the equipment and some, including Robert, went ahead by rail. Unfortunately, the truck broke down and never arrived. The rest of the group eventually arrived by train, bringing what equipment they could – which did not include Robert's meteorological equipment. He was stuck in the south of France with no work to do. It seemed to suit him just fine, and he enjoyed going to the village and meeting the locals.

After six weeks, however, the group was left with very little money for the homeward journey. Robert and a friend decided to hitchhike via Italy. From Turin they travelled through the Brenner Pass to Austria, where they slept under a hedge, then over the Alps to Innsbruck and into Germany. From Augsburg the pair got a lift to Munich where they split up. Among the experiences Robert would never forget were sleeping in a seemingly quiet churchyard where the church turned out to be a cinema full of patrons that flooded out around the duo, being picked up by a creepy man who wanted to take him to a hotel on the Rhine, staying in a hostel in Cologne that had been the dungeon of a former police station, and having a woman throw chocolate to him from a window in Aachen. It was a memorable trip.

Robert passed his university year and moved to the agricultural campus at Sutton Bonington, where he shared a room in the hostel with a young man named David Charles. As he had in his first year, Robert threw himself into the social life, took a lead role in a stage production and continued with his rowing. The result: Robert failed his second year. David had been warning him about this. Robert found himself sitting by the railway lines contemplating suicide. However, a concession was made due to his significant input to the rowing team and the ballroom dancing classes he had organised, among other activities: Robert would be allowed to re-sit his exams. He spent the remainder of the summer studying – which was a pity, as he'd been picked to spend the summer working in the gardens at Versailles. Instead, he went to Imperial Chemical Industries (of all places), a research centre in Fernhurst, Surrey. It was a huge blow.

Observing that David Charles was achieving quite well, Robert asked him to share accommodation again and to help him to keep on track. The pair got a student house together, and Robert's second year at the Sutton Bonington campus commenced. He worked hard, won the Ashgate Prize for horticulture and was

CLOCKWISE FROM TOP LEFT: *David, Madge and Bill, Whitecliff Bay, early to mid-1950s; Bob, Devizes Grammar, c.1955; Studying by night in the south of France, 1959; Bob and friends, Les Alpes-Maritimes, 1959; Bob (third from right) on stage in* The Prodigious Snob, *c.1958.*

admitted into the honours stream. In his final year he completed his honours thesis on the Latin American medicinal plant *Cephaelis ipecacuanha* – traditionally used as an expectorant and an emetic. The study involved propagation work using growth chambers.[59]

David's nickname for Robert was the unusual epithet 'Gascoin'. When Robert queried this, David replied: 'Well, the gas metre takes a bob, and you are a Bob, therefore, you are a gas coin.'[60] Robert hadn't thought of himself as Bob before, but he liked it. He began signing himself as 'Bob' in early 1960. Madge defiantly persisted with 'Robert' for the rest of her life.[61]

Now 21, Bob had moved far from home and changed his name. These were important steps to asserting his own identity, but in his personal life he felt increasingly that he was becoming an actor. He knew he was attracted to men and had what he called 'some amazing crushes' at university. But he kept these hidden, in part by being very sporty (as he had at secondary school) and in part by always having a girlfriend. One of these was Wendy, whom Bob met through a mutual friend, Danny. 'There was a genuine feeling of attraction when I first met Wendy, and it so happened that she was teaching at Loughborough, which is close to Nottingham. She had a little scooter, a Puch, which she lent me … And we got on very well.' Wendy even drew one of the illustrations for Bob's thesis. But the relationship was not going to go anywhere and, in the end, she married Danny.

Bob's crushes on male friends continued but were barely allowed expression. A group of gay students met in one of the common rooms. 'I sometimes walked past to see them, but I didn't have the guts to join in. I was a macho man, you see … I think it's why I wanted to be a long way from home, because I felt that I was different and I didn't want to expose my homosexuality.' He was not consciously aware of this motivating force – at least, not all of the time. Perhaps it was part of that sublimation process he later identified in himself.[62]

The popular song 'Moon River', sung by Audrey Hepburn in the 1961 film *Breakfast at Tiffany's* and covered by Andy Williams in 1962, became Bob's theme tune. A drifter, off to see the world, and the words 'there's such a lot of world to see' seemed to fit. He began to formulate a plan to move far from home – much further than the distance between Devizes and Nottingham.

In July 1962 he set off with friends for the Austrian Alps, hitchhiking from Ostend to Salzburg.[63] After sweltering in Innsbruck, where he had visited in 1959, he climbed Habicht in Tyrol to see the view from the top, but descended in mist. He wrote to his parents, 'It was most exciting crossing over snow fields and very rough country and the mists made it look pretty grim.'[64] On 22 July he climbed Kitzsteinhorn.[65] 'Words cannot describe the beauty of the mountains,' he told his

ABOVE: *Bob with Wendy and friends, c.1961.*
BELOW: *Graduation drinks, 1962 (Wendy on right, David Charles centre).*

Graduation, 1962, with David Charles.

Uncle Jack, 'sky and snow all seen under perfect alpine conditions of clear sun and sky'.[66] He remained on the mountain for two days before returning to Innsbruck.[67]

Bob graduated from Nottingham in July 1962 and went home to Devizes, where he picked up work in a mill and again experienced extreme back pain. While there he was offered a job with the Department of Agriculture in New Zealand, to train as an advisor in horticulture. He knew about the South Pacific nation from his natural history and geography studies, and the idea of emigrating appealed. In addition, as he saw with hindsight, New Zealand was attractive 'because it would be a break with family ties, and it wouldn't matter so much if I was found out'.[68]

In December 1962 he departed for New Zealand.

Bob, Mt Eden, Auckland, 1963. (Photo courtesy of David Dennis)

2: Auckland and Pukekohe, 1962–66

Bob Crowder began his 'epic adventure to New Zealand' on 4 December 1962, one month before his twenty-fourth birthday.[1] His journal, previously dedicated to weather reports, immediately became more poetic. 'The commencement of the trip to the sun started as it should in truly severe winter weather,' he wrote from his cabin that night. 'Roads were chaotic as freezing fog froze to them creating ice rink conditions … the silver birches festooned with rime and gleaming against a rich blue sky, a truly beautiful farewell by England after a year which has been notable for its ice-cold bleak habit.'[2]

Madge and Bill drove Bob to the port. It was a sad occasion. 'It was awful to see you go,' Madge wrote the next day. 'The next time I see the dock I hope it will be to see you walk off the boat, I think I would really enjoy that … goodbyes are not very nice.'[3] Bill reiterated her thoughts:

> [A]t the time of writing we are still suffering from the casting off of the ropes … I don't think the awful truth burst on us really until that moment. Your mother is doing the stiff upper lip business nobly, but I notice she has not been able to straighten out your bed yet … trouble is you have been such a busy … man that we are surrounded by your goings on. The garden, hedge, stone walls, stepping stones etc, too many to enumerate. At the moment all these things have a very disturbing effect on your mother and me.[4]

Bob shared a cabin on the SS *Southern Cross* with three others: a married man heading to Levin to do irrigation research, a 'dour northerner' he found difficult to talk to, and an 'active alert youth from Scotland' on his way to Wellington. 'Being the youngest of these three he and I do seem to have more in common and hence this is the first friendship made.'[5] Dinner on the first evening brought another cast of characters: Velma and Margaret from Melbourne, Roger from Christchurch and an Australian man who helped to keep the conversation going. The ship's orchestra was 'pathetic' and the violinist, in particular, 'ludicrous'. Never mind: 'The tavern as the local will prove to be the place for the lads to meet.'[6] It was true: there was dancing and drinking every night, although Bob felt the ship was 'not satisfactory for youth, too many old people blight the air and the young find themselves in the minority'.[7]

After five days he met Rod and Ted from Western Australia, 'really bright sparks'. Rod was 'obviously the dominant partner'.[8] They may have been the first gay couple he'd met, although he didn't write this in so many words. As more people joined the group, Bob pondered his options regarding the women, noting that he was 'loath to settle down with any one bird at this stage as they are all so pleasant in this particular group'.[9]

'Sun-worshipping' was a favourite pastime, and Bob observed, 'Roasting bodies, hot, sweaty, and often ungainly with rolls of fat, but again often beautiful in physical perfection.' Certain 'antics' by crew members on the foredeck gave rise to the 'question of homo-sexuality'. A few people had already pointed out that some of the stewards were 'queers', 'and I've no doubt there are some as in most societies, no doubt there are plenty among the passengers also'. He voiced his frustration:

> The whole approach appears wrong to me and the situation, if it exists, is no worse than the immoral attitudes and postures of certain couples of ♂ and ♀. A highly sexed person uses prostitutes or people unable to resist the high sex appeal of such persons, the homo-sexual has equally strong emotions and if no corruption of innocent persons is involved, I see no reason why his needs should not be recognized as equally 'natural' to those of prostitutes. Both aspects of immorality or vice are equally to be abhorred and it is interesting to find that those who claim to be most broad minded on the subject of prostitution are generally those exhibiting the least compassion or understanding and the greatest distaste for the act. It is not easy to know how oneself would react to an approach of either of these vices and it is as well to keep from sneering and prejudging on either case, a sick mind is more to be pitied than a sick body.[10]

Bob's back pain had returned with 'severity', and he experienced 'fits of depression' that he believed would 'continue until a final reckoning is obtained':

> It is my inability to transmit myself that is at fault. Despite a great love of people, I am unable to give the impression, it is the same all through my life ... Such is the effect of another Tropical day, 82F on the bridge and o'so humid. Not at all pleasant and extensive cloud continued to blot the sun out.[11]

He perked up in Trinidad during a day of sightseeing: 'The beauty and joy of the day cannot be captured in words, only in memory, with gratitude for good companions to share it with.' At a bar in the Port of Spain, however, he was exposed to 'the whole of perversity from the worst possible homosexual angle to extreme prostitution'. Some of the stewards took part in beauty contests, wearing makeup to look like women, while 'the girl dancers spared no punches in the contortions of the flesh and expression ... The whole show was pandering to certain tastes ... As for the negro hostesses, the furtive hand grasping and bottom pinching were not my line of country but it certainly served as an eye opener.'[12] At Curaçao, 'Things were said in the presence of birds that would never be dreamed of in England. Rod was positively disgusting by English standards, boy it is going to be difficult to adjust if this is an example.'[13] His frame of reference seemed to be falling apart. 'I'm not sure I want to adjust to that. In fact I know I don't. Still, never pass judgement ...'[14] Then again, he didn't go to church because 'a doubt existed as to my behaviour, it is important that I should not slip into the loose thinking and talking of my friends'.[15] On 17 December Bob and his friends arrived in Panama. He was fascinated by the lock system in the canal and the efficiency of the US operation there. When his friends stopped for hamburgers, he strolled on alone. He acknowledged his wanderings were 'due to considerable disturbances which prove a worry and strong urge'. His caginess about this was explicit – 'I do not commit my deepest thoughts to paper' – but it is not difficult to appreciate what was going on:

> I was excited when I spied through the corner of my eye a figure making for me. 'Have you a light please', it was a coloured boy in squared shirt, very neat. I replied in the affirmative and moved in round the square, past many young people on the seats, convinced I was to be approached. He did not follow so I walked back to the step tops to stand and gaze at the busy scene below. Why? I only know I knew I would be approached again and wanted to see what the outcome would be.
>
> We were soon in sporadic conversation and in answer to my question of 'and what do you really want', he said 'Well, I expect you will laugh, but I want you.' ... To say that I was not disturbed by this would be a lie, but I was not disturbed to the point of revulsion ... In this case he was a normal decent fellow, very friendly, to the point and quite honestly, if I had not got all my good friends waiting for me, I should have spent a good evening with him, though I trust he would have had a disappointing time. This is the first time anything like this has happened to me and it would have been good to know why he chose me of all the people there on the place, why my lack of money did not deter him. How he became what he is, it is only by talking and trying to understand other people that we can appreciate their troubles and difficulties fully.[16]

After this encounter Bob returned to recording the weather and noting seabirds. He broke away from his group of friends and made the acquaintance of two lads from New Zealand who were 'willing to talk on more serious topics than the Rod, Ted and girls group'. He found the party scene taxing: 'The atmosphere of enjoying oneself by making as much noise and disturbance as possible, [is] artificial and terrible to observe.'[17]

In Tahiti he convinced his friends to tour the island instead of going shopping, which, he said, 'bores me intensely all this waste of time'. The experience 'was a typical example that a group though very pleasant has many drawbacks and lone wandering can be more rewarding'. He admired the 'truly magnificent' colours of the island, the bougainvillea, hibiscus, 'the yellow and whites, creams and greens all swelled into loud exotic splendour under a blue sky and the many palms waving in a fresh trade breeze'.[18] He enjoyed a swim at a 'black sweltering beach where the water was warm and lush and the waves hard and pounding'.[19]

In Fiji he gave his friends the slip altogether and visited a research station where he learned about the work undertaken there with cows, pigs and poultry, and rice, taro, coffee, tapioca and bananas. He went to the Botanic Gardens to brush up on plant identification. During his wanderings he met Jeremiah, an 18-year-old Fijian from an outer island who was studying at the Methodist Mission school. They walked through the bush together and met one of Jeremiah's friends, who gathered green coconuts for them. Bob visited their hostels, inspected their crops and dug up tapioca to eat, and finished his visit with orangeade and ice cream. This was more like the kind of travel he enjoyed, and he reflected, 'I shall remember Fiji with the happiest of memories of this whole trip'.[20]

The voyage was nearly over. The following day Bob turned 24, and three days later – on 7 January 1963 – the ship sailed into Wellington Harbour. The weather was foggy and damp: 'Wellington was a grim city in this weather, very ancient and a little decrepit and forlorn.' He was met by people from the Department of Agriculture and taken to the office, which was in a condemned building 'and not very beautiful … The first day was therefore rather depressing and grey and not the NZ we had expected.'[21]

The journey to Auckland by train was discouraging at first for the young adventurer. The sky was overcast and he found the landscape oppressive and bleak. But once past Hamilton his spirits lifted: 'Auckland is at least a city and looks like it.'[22] Here were sunny, clear skies and magnificent but as yet unknown vegetation.

The Department of Agriculture sent David Dennis to the railway station to collect Bob. David had joined the team two months earlier and was a few months older than Bob; there was some thought that the two would get on well together. For David, at least, that was an understatement. 'I felt a sort of electricity,' he later recalled.[23] David helped Bob to search for accommodation in the city. One man offered to share a house for just £4 a week including food. Bob and David were 'both thinking the same thing, what does he really want, homosexuality? It can't be money or anything like that.'[24]

Not having somewhere to call home provoked a pang of homesickness. Bob's parents sent messages of consolation: 'You will feel much happier when you have your own little home to return to when work is finished. I can understand you feeling worried & a bit homesick when you are adrift with nowhere to spread your belongings and relax,' Madge wrote.[25] Bill took the opportunity to share a rare insight into his time at war: 'Don't worry about being a little sorry for yourself at the moment, it's very natural and nothing to be ashamed of ... I remember I had the same trouble when I went to India despite the fact that I had my crew, and it continued on and off until I came back.'[26]

Bob made friends gradually and tried to be sociable, but it did not always go according to plan. On one occasion he arrived at his friend Eileen's house to go dancing with her but was confronted instead by an aggressive, overprotective brother. Bob walked off without seeing Eileen – then 'ran into a young lad with homosexual tendencies and had a pleasant evening with him, not, I hasten to add, in any sense other than a talk over coffee. Why I'm so plagued with this now I do not know.'[27]

Meanwhile, David showed Bob the dance halls and a coffee bar, which Bob thought 'proved very spiv'.[28] The two got on well and by late January were flatting together on Wairakei Road, off Mount Eden Road. The flat was owned by the father of one of David's Massey friends. 'So, your new friend is another David,' Bill Crowder wrote to Bob. 'You seem to favour that name, and it must be a big help to have a friend who has connections.'[29] David was becoming increasingly 'besotted' with Bob, who seemed oblivious to his feelings. Bob later learned that David was wrestling with his sexuality and voluntarily seeing a psychoanalyst to try to become heterosexual. 'In the end [the counselling] resolved a conflict. I was a gay man,' David concluded.[30] In Devizes, a Christmas card had appeared from Bob's ex-girlfriend Wendy. Bill gently observed that Madge had not replied to it, 'didn't want to do the wrong thing so to speak, but if you write to [Wendy] as no doubt you do, tell her the position, if you like say we've lost the address or something ...' There were changes at home – Madge was having Bob's room redecorated: 'I fancy you are going to have rose sprays on one wall at least,' Bill wrote – and David's engagement had broken off. 'I can't say I blame her,' remarked Madge, 'as David doesn't seem the least bit interested in getting married.'[31]

But the real news, with which Madge's correspondence in particular was always loaded, related to the weather and the garden. The weather had been the worst on record since 1830. But when March arrived Bill reported that the broad beans were up, although the ground was still too frozen to harvest parsnips. A fortnight later Madge informed Bob:

TOP: *Bob's Christmas card to his parents in 1965 featured a photo of him on Mt Taranaki.* BOTTOM: *Bob with friends, Auckland, 1963. (Photo courtesy of David Dennis)*

Everything is shooting up in the garden it's amazing, all the crocus on the front lawns are out (those you put in) lovely clumps on the back lawn & snowdrops & those [illegible] mauve flowers on long stalks are beginning to come into bloom. We have the potatoes upstairs laid out to sprout, & hope to get some in as soon as possible. The parsnips are still perfect, we had some this week also I am still managing to salvage a few sprouts.[32]

By May the garden was in full swing: 'The aubretia blossoms are large & flawless & the Siberian wallflowers look very healthy, so do the Canterbury bells & the sweet Williams which have survived. The blackcurrants are covered with blossom, if all goes well we should get a bumper crop. The strawberries are beginning to flower.'[33]

It was some consolation to Bob to know that his garden was thriving. No doubt he was also interested in the weather updates, at which Madge was particularly adept. She filled dozens of letters with detailed accounts, such as this example:

The met. men really stuck their necks out on Weds telling us there was a lovely high pressure system coming in & the next few days would be fine all over Britain with light Southerly breezes & temps above the average. That did it, as usual, the next day it went cool & cloudy & apparently the large area of H.P. [high pressure] had halted off N. Ireland letting in some more cold air from the North. Today it has been quite cold again with sunshine & showers even a few snowflakes when it rained … They had thunderstorms around the S.E. yesterday too.[34]

Madge also liked to dispense motherly advice and was concerned about Bob's back problems: 'Robert do take things more carefully … Don't go doing too much rowing or this skiing lark which is rather noted for causing injuries, remember you have your career to think of.' She urged him to 'have one or two evenings in the week when you can relax and have an early night so that you are kept fresh and alert for your day's work.'[35]

Whether early nights were the answer or not may be debated. Bob and David Dennis shared a room, and eventually the sexual tension David felt became too much. After six months he appeared one night at Bob's bedside, prompting Bob to move out in June.[36]

Bob took a flat with Keith, a gay man in Parnell, and found himself in a network of gay men and illicit liaisons. Among the characters he met was the son of the dean of the cathedral, and his new neighbour was gay as well. 'I knew I wanted to join in, but I didn't,' Bob recalled.[37]

David's feelings continued to grow and somehow a sexual relationship – risky at the time – developed between the two. 'We were in the public service, and it was illegal … That is really when things crystallised for me regarding the gay world.'[38]

Bob and David were both members of the Alpine Sports Club, which organised year-round group tramping activities. After a party one night David suggested to Bob that they drive to Titirangi, where the club had a tramping hut, 'and it all happened out there'.[39] According to Bob, a cave near Piha was the first place they had sex.[40] David recalled, 'I don't think [Bob] ever felt an intense attraction towards me, [yet] he didn't seem to dissuade me.'[41]

Bob thought it best to remain single. 'Opinion of girls still not very much changed and getting more & more convinced that self-reliance and non-dependence is the way to life,' he wrote.[42] Instead, he immersed himself in sport, explored New Zealand and socialised. He took up basketball, which brought on another serious bout of back pain, but this didn't stop him being active in the Alpine Club and taking part in adventures such as skiing by moonlight and climbing Rangitoto, Ruapehu and Tongariro. He enjoyed the days of 'breathtaking' views and evenings of 'fire, dinner & sing song'. The most dramatic of these adventures involved getting lost in the fog on Ruapehu in August 1964, when he and another club member were saved by a Mountain Rescue team.[43]

Bob took every opportunity to hitchhike around the country, sightseeing and learning the ecology of the land. During a walk through the Waitākere Ranges, he was

> overawed by the fact that I was a British lad that knew a lot about the natural history of Great Britain and was used to knowing where everything was, and could recognise a vast range of trees, shrubs, herbs, grasses in Britain, and being emptied into Auckland was a gobsmacking experience, because suddenly I didn't recognise anything. And walking through the Waitakeres there was nothing I could recognise at all, and that was quite traumatic for a natural-leaning person like myself.[44]

His answer was, typically, to study furiously, to learn the common and Latin names of plants, write extensive lists and commit it all to memory.

Bob kept in touch with friends and family back home. He wrote to his grandmother Elizabeth Crowder to wish her a happy birthday (she turned 90 in May 1963, as Madge reminded him), and to Mrs Belben, whose garden had inspired him as a boy. He wrote to his father in April, enclosing what was apparently a large birthday gift of money. Madge was overcome: 'I don't know what to say … it really gave us a lovely warm feeling and made everything seem worthwhile even though it has meant us losing you.' As for Bill, the cheque took his breath away: 'Your munificence was really rather alarming … thank you most sincerely for a gift which I realise, knowing you a wee bit, is perhaps something you have wanted to do for a long time. Both your Mother and I would willingly give it back and add some more to see your cheery face rushing down the road again, at least never forget that.'[45]

Bob enjoying the Auckland outdoors in 1963: Taking in the view from One Tree Hill (above) and swimming at Point Chevalier (left).

(Photos courtesy of David Dennis)

Like Madge, Bill also liked to keep his son appraised of the garden's progress in his characteristic chatty way:

> I have also had my first full row of peas pop up successfully. The first for a long time, which is surprising since I just shoved them in with little preparation and no red lead. The spring cabbage you planted has had a shaky winter but I transplanted what had survived and made two good rows which are now coming along fine. The strawberry plants are now looking very well after a good forking over with fertilizer, and the potatoes are in though not through. We have our usual 50,000 sprout and cauliflower seedlings and I suppose shortly we shall start all over again with millions of wallflowers etc and plenty of backaches transplanting them.[46]

Bob discovered Auckland's social scene. 'There was a great coffee culture in Auckland in the 1960s. It was impressive ... a lot of the coffee bars were open right into the night, with little trios playing ... Somebody would have brandy or whiskey in the hipflask and we would lace our coffees. And there was really good music, and dancing.' The jazz scene was very much Bob's style. His favourite venues were the Hi Diddle Griddle on Karangahape Road and the Montmartre and Lautrec coffee lounges on Lorne Street (close to where Auckland City Library is now), where the Mike Walker Trio (the resident act at Montmartre) and Mike Perjanik Band were regular jazz acts.[47] He was sometimes plagued by regrets after social occasions: 'I vow not to attend any more parties. What a waste of time and intellect. Nothing achieved, nothing gained – only infamy and a wasted Sunday morning in bed. Got home 5am. May I remember these words at a later date.' A month later, after another party, he commented, 'My vows all to naught, still enjoyed it all.' A week later he attended another 'fab party ... really let my hair down. How I shall face the office tomorrow I don't really know.'[48]

Bob was also active in church life. On 12 April 1964 he recorded, 'Attended church and accepted position of chairman to Young Anglican & Youth Committee.' He took an active role in this. Youth group on Wednesday evenings might involve judo, basketball, table tennis or dancing. In June he organised a youth festival and made specific mention of the individuals who helped to pull the event together. The youth group also contributed to his habit of staying up late, for example: 'taking Russel to meet Jean-Pierre a French boy Judo expert'. He then had a long talk with Russel afterwards. 'Late night again.' Bob's interest in making connections between people had not abated.[49]

Among other church activities, he attended discussions with the Takapuna Methodists and the Knox Presbyterians on forms of worship, and with Father Murray of the Catholic Church, who intimated that 'there is little hope for unity of the churches'. He continued to struggle with his own feelings of judgementalism.

Bob outside the Department of Agriculture building, 1963.

One individual he described as 'slutty' and 'the most self-loving & selfish person ever. GOD GIVE ME STRENGTH. If I've ever despised anybody it is him, poor fellow. I ought to have compassion he is miserable & alone.'[50]

Bob seemed to be searching for something. In May 1964 he read J.K. McCarthy's autobiography *Patrol into Yesterday*. McCarthy was a government officer and soldier and, as Bob recorded in his effusive reflections, director of the Department of Native Affairs in Papua New Guinea and a member of the legislative council there. He was particularly remembered for his heroic rescue of around 200 people after the Japanese invasion of Rabaul in Papua New Guinea in 1942, for which he was awarded an MBE.[51] Bob was moved: 'a deep impression has been made on me'.[52] McCarthy's experiences resonated with Bob's ideals of heroism, leadership, and a sense of calling.

Lastly, there was work, or, as he referred to it, 'work work & more work'. His diary increasingly mentioned his dissatisfaction: 'Work not too good today, very tired'; 'Feel apathetic & tired'; 'A wretched day, need a change a lot.'[53] Bob's mentor in his role was Bruce Coleman, 'who had a great rapport with the growers. He gave me the best advice of anybody when I first arrived ... He said "come out with me. You'll learn a lot but don't open your mouth for a year."[54] Bob soaked up these experiences and came to know many growers in the Auckland area (including the Goks, now-famous Chinese market gardeners whose work on kūmara production with Bruce Coleman helped secure the kūmara industry in the face of a black rot break out). Many of these connections later played a role in Bob's teaching work at Lincoln.[55]

The advisory work was an extension service:

> We were servicing growers around Auckland, and we'd go out and advise on whatever the problem was ... Auckland was in flux and a lot of good soil was being taken, and some of the growers didn't know what to do, and Bruce Coleman advised that they should capitalise on the land and reinvest the money and he would tell them good areas to go to. And quite a number of the major vegetable growers in Auckland went north of the city, into Waimauku and so on, on his recommendation, and set up very good holdings up there. Very good learning for me.[56]

In addition to his exposure to the horticultural scene in and around Auckland, Bob picked up some practical skills. The department had a 'very good apple specialist' who took some of the staff out to topwork or graft trees in orchards. Armed with his new skills, Bob 'bought a little scooter and ... went around pruning apple trees around Auckland in the weekends. We were encouraged to do things like that.'[57] This could take up an entire day. On 20 June 1964, for example, he did six hours of pruning with one customer (and received an excellent dinner), then set off to another, where he noted that his work the previous season had resulted in an abundant crop of pears and apples. The following day there was more pruning work. By the end of the year he had 31 clients.[58]

On one of his walks through the Waitākere Ranges, Bob took a shortcut across some fields and ran into Peter and Margaret de Waal, whose property it was. They invited Bob inside for dinner, 'and that was the beginning of a really good relationship for two or three years. We ended up planting tamarillos'. The tamarillos didn't thrive – 'I think they got mildew' – but the friendship did. 'Peter was a South African, and we used to have long chats about fascism and socialism and everything. I used to go and stop up there on weekends, to get the work done. They were very sure that there would be problems in South Africa, and it was moving into the Vietnam conflict.'[59] Bob was beginning to realise that he had a political identity. He remarked in August 1964 that he had been reading a book about Cuba that was 'illuminating & idealistic' and talked about its 'magnificent progress'.[60] Reflecting on this period in 1966, Bob wrote, 'Peter is quite a remarkable person

ABOVE: *Bob (on right) with some of his youth group, Pukekohe.*
RIGHT: *Bob in the garden at his Pukekohe residence, c.1965.* (Photo courtesy of David Dennis).

and I owe him quite a lot for making me think and seek out the truth on the issues of the day.'[61] It was a pivotal moment in which Bob discovered something new about himself.

Part of this developing identity was his aversion to bureaucracy, in which he was mired at the department. It wasn't so bad while he was in Auckland, but in 1965 he was posted to Pukekohe as the sole advisor, a position he held for two years. It was a big step for a young man – he was still only 26 – and indicates the confidence his superiors had in him. He had already worked with some growers in Pukekohe; among other things, he had undertaken an onion survey there in July 1964.[62] He found a place to live and settled into his post.

The red tape soon started to rub: 'I had some run-ins with some of the officials.' On one occasion Bob discovered that one of the major growers had dipped his pumpkins in Baylotan, a 'totally illegal' mercuric insecticide. Bob impounded the pumpkins immediately, an approach that was 'against protocol'. He knew, however, that if due process had been followed, the grower would simply have taken the pumpkins to market and sold them. As for due process, 'I didn't warm to it.'[63]

As usual, Bob became involved in social activities. He helped to start a youth group in Pukekohe based on 'a sympathetic coffee bar'. The youth group was a success, particularly in organising a dance for the town's centennial celebrations. A number of acts performed, including Larry's Rebels and Ray Colombus and the Invaders, and some 600 or 700 people attended. The dance was 'a huge success … a total sellout' and was the only centennial event to return a profit. 'Somehow there was an affinity between myself and young people, and I could always rely on them to help with organising.'[64]

One area of his life that did not continue was his involvement in the Church. 'I didn't like the Anglican church in Pukekohe. I had been the chairperson of the young Anglicans at St Mary's cathedral [in Auckland]. When I arrived in Pukekohe I thought I'd have a look. I didn't enjoy it, didn't like the atmosphere.'[65] It was the end of Bob's church involvement. His faith seemed to transmute, somehow, and became like that of his mother: strong, only vaguely codified and free of any formal religious structure.

He found Pukekohe 'a strange place'.[66] There was palpable racism present: 'The Indians were a level above the Māori, and the Chinese a level above the Indians.' Pākehā would not go to the same bars as local Māori. A number of Māori dressed in shorts and gumboots came to one of the formal dances Bob organised. This did not go down well with some of the Pākehā attendees 'and it was pretty tense there for a while.' For Bob, an egalitarian at heart, such scenes left a bitter taste in his mouth: why could people not see past their own biases and look with open curiosity into the lives of others?[67]

Professionally, Bob was making an impact. His name was known through his good work in the department and at the end of 1965 he published his first article, on witloof, in the department's *Journal of Agriculture*. It was evidence of his early interest in unusual vegetables that might broaden the Kiwi palate. He was intrigued, even though 'the time and labour involved prevents fortunes being made from the crop at present'.[68] He also wrote a piece on the potential commercial uses of New Zealand native plants. The department head rejected this article, however, on the basis that Bob was new to New Zealand and couldn't know anything about the indigenous vegetation.

In February 1966 Bob confessed to his parents that he was 'a bit fed up with the whole running of the Dept. Agric.'. So much so, in fact, that he had handed in his resignation. There was good news, however: 'Without even applying for a job I got two very good offers which I followed up over the Christmas holiday.' Bob had been shoulder-tapped by Professor 'Mac' Morrison of Lincoln College, who was establishing a new Department of Horticulture there. He flew to Christchurch for an interview at Lincoln (where he caused shock among the academics by arriving in shorts) and 'accepted an offer as lecturer in horticulture with special reference to vegetable production'. He also registered to do a PhD and told his parents, 'This is quite a boost for me of course. Lincoln is very well known overseas for its agriculture and as a result a PhD from here will be quite valuable.'[69]

One of the prominent growers in Pukekohe, Henry Wilcox, proved to be 'instrumental, at a formative level' in Bob's thinking about his PhD.[70] Henry showed Bob the Stanhay Drill he had just purchased, and suggested Bob do his PhD on that. The Stanhay seed drill was absolutely new and stood at the gateway to intensified, industrialised precision agriculture. Bob was enthusiastic and his subsequent work on precision seed drilling shaped his entire career.

The other job offer Bob received was from Massey University, which was also setting up a Department of Horticulture. The memory of his dreary train ride through the North Island after arriving in New Zealand, and his particular dislike of Palmerston North, certainly influenced his decision to accept the position at Lincoln.

Bob was now 27 years old and excited to be shifting out of government bureaucracy, but even more so by the call of the mountains. This was pressed home over the summer on a climbing trip in the South Island that involved energetic adventures in snow, sleet and heavy rain, difficult river crossings and complicated rope work. Sir Edmund Hillary was with the party. 'I shall be sorry in some ways to leave the soft warmth of Auckland for the more English climate of Christchurch but I feel badly in need of a change at present. Christchurch is of course wonderfully situated for skiing, skating and mountain climbing. The Canterbury ski fields are only 1½ hours away and ice-skating is even closer.'[71] Lincoln even had a ski hut in the Craigieburn Range, which helped to seal the deal.

On 8 April 1966, Bob set off to drive south in his Humber 90. It was, he wrote, the 'beginning of the great trip down. A glorious morning of brilliant sunshine and on it continued until the end of the day.'[72]

The Horticultural Research Area, Lincoln, c.1966.

3: Lincoln and the Horticultural Research Area, 1966–76

Bob began work at Lincoln College in April 1966. In his diary he refrained from commenting on the fact that he was establishing a new life, mentioning only that 'some strato-cumulus were present especially further to the North over ChCh'. He pondered the climatological effect of Banks Peninsula, which broke up the cloud cover, leaving the skies clear over Lincoln.[1]

Lincoln College was originally established as the School of Agriculture in 1878. It was renamed Canterbury Agricultural College and then Lincoln College in 1962, a few years before Bob arrived. Lincoln was the agricultural campus of the University of Canterbury (it became independent in 1990). In 1966 the college had around 600 students and roughly half of these were resident on the campus. Some 12 percent were studying horticulture; the rest were mostly agricultural students, and the majority came from farming backgrounds. Most were male and in their late teens or early twenties.

Bob moved into his on-campus accommodation in Hudson Hall and got into the routine of delivering lectures. Despite some frustration with the length of time it took to prepare these ('I'm not used to sitting down for long periods'), he was pleased with how they were received.[2]

He wasted little time in getting to know his surroundings, particularly the mountains. One week after his arrival he was off to the Arrowsmith Range, and in May he joined a working party at the Craigieburn ski field. When winter arrived he made numerous trips to Lake Ida for ice-skating, and in August he skied at Craigieburn – the snow was in 'excellent condition' – and at Round Hill near Tekapo: 'brilliant day … with views of unsurpassed majesty over the Cook region'. He skied at Craigieburn again in September and visited Oaro, near Kaikōura, where he spent a day collecting plants and walking around Kaikōura peninsula to visit the seal colony. 'The sea [was] a marvellous milky blue & very beautifully surfy,' he wrote. In October he made several trips to Arthur's Pass and different ski fields, and camped by the Hāpuku River in the Seaward Kaikōura Range.[3]

Trips into the backcountry were an important part of Bob's life for many years and compensated somewhat for life on campus. Student revelry there centred on drinking, which did not appeal to him. He was 'horrified with the level of the social life at Lincoln'. The height of entertainment was a game where participants had to pass a balloon, held between chest and chin, around a circle without using their hands. 'They weren't the kind of social functions that I would want to go to.'[4] Taking matters into his own hands, he began to organise social events on campus. He chaired a student debating event in 1966, and his Christmas Eve that year was 'socially the best ever – a remarkable party – Champagne – terrific company & obviously on level at last. Ended 4am after midnight Mass. Memorable.'[5] The following year he organised a staff Christmas party.

Bob began to develop a social life off campus as well, and discovered The Landing restaurant in Cashel Street and its owner, Leon Langley.[6] A typical Friday evening saw Bob at The Landing for dinner, and from there he would progress to an upstairs bar at the United Services Hotel, where he met and befriended the band The Village Gate. When the bar closed at 10pm, Bob and his friends would cross Cathedral Square to continue their evening at a coffee bar.

The establishment of horticulture as a scientific discipline at Lincoln and Massey in 1966 was a significant moment in the history of horticulture in New Zealand. It was in part driven by the country's need to diversify away from meat, wool and milk, and to be competitive in an international environment that was shifting rapidly towards mechanisation, industrialisation and precision. Wilcox's Stanhay Drill was just the beginning of a massive effort to bring New Zealand's primary sector up to date. Crops of a uniform size and standard were wanted, since uniformity made packaging easier, and in the dawning era of supermarkets, this was becoming important.[7]

Mac Morrison provided an overview of the new horticultural programme at Lincoln in 1967, in which he stressed that horticulture was more than just gardening: it provided opportunities for economists, engineers, chemists and even artists to apply their skills in practical ways. Horticulture, he wrote, 'takes design and form of structures and uses them to ensure that our landscape, and the construction upon it, is as it should and can be – aesthetically pleasing'.[8] His ideas echoed Bob's own sentiments, even if Bob had not yet put these in so many words.

In 1969 Morrison elaborated. Horticulture, he said,

> is more correctly defined and practised as a further, more intensive stage of land use than traditional agriculture. In fact as land values rise farmers must turn to high return crops. For this change to take place the farmer must be given low-labour techniques, protection from wild fluctuations and sound management advice. This fairly well defines the aims of the present Department of Horticulture at Lincoln College … With this general aim of introduction of intensive cropping systems into New Zealand agriculture go the detailed scientific investigations of changes in microclimate, water and fertiliser use, and morphological physiology. This defines the department's scientific programme.[9]

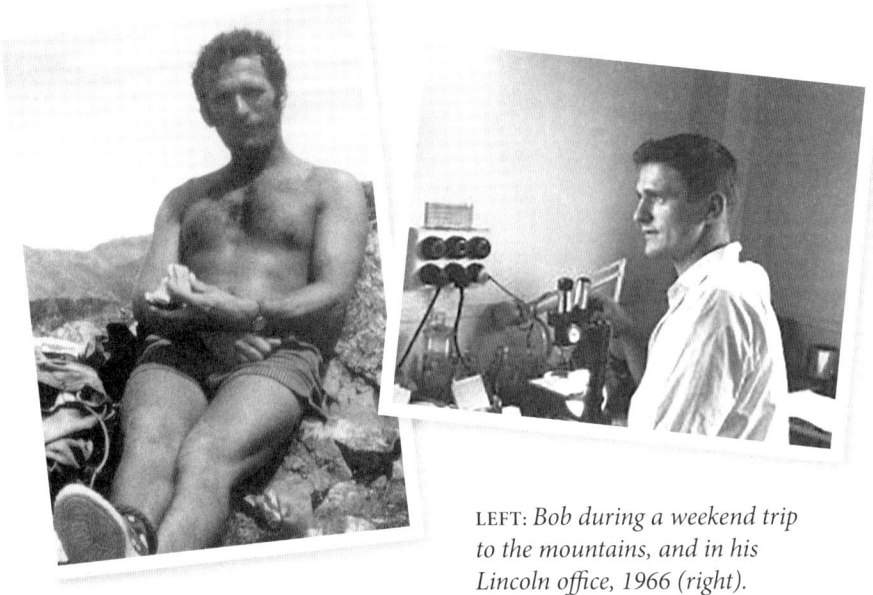

LEFT: *Bob during a weekend trip to the mountains, and in his Lincoln office, 1966 (right).*

Alongside duties as a lecturer in vegetable production, Bob's role included research on the industrial 'development of high crop density in extensive production of vegetables'.[10] A Horticultural Research Area was established to support the development of new horticultural crops and to explore and experiment with innovations in growing and harvesting techniques. It was by far the largest 'garden' Bob had ever been fully committed to. It required his constant attention, as his field notes show. Planting, weed and pest control and irrigation trials were all well underway by late 1967 (with, among other things, a spray of Dieldrin for thrips on the onions. Dieldrin, an organochloride chemical, would later become a banned substance). As well as looking after vegetables (tomatoes, onions, cabbages, carrots, lettuces and peas), he had some involvement in other areas, such as irrigating trees. In January 1968 he made notes on fruit crops: strawberries, raspberries, boysenberries, blackcurrants and an orchard. A year later parsnips, celery and Brussels sprouts were added to the mix. Numerous pesticides and herbicides were pressed into service here during 1968, including Dieldrix, Linuron, Ciba, Lasso (a Monsanto product), Paraquat, Diphenamid, Treflan, Benefin and Malathion.[11]

In 1968 Bob received assistance in development of the unit and its research programme from a young Englishman, Bob Douglas, who had just arrived in the country. The Horticultural Research Area land was very exposed to the weather – 'you could see the Alps; the wind just used to howl' – and Douglas worked to establish shelter belts, cutting poplar poles from existing trees around the campus and planting them out on the weekends.[12] It was hard land to grow trees in; after particularly dry conditions in November that year, Bob recorded, 'irrigation [was] put into operation to save poplar trees which were thought to be established'.[13] Douglas worked there for eight or nine months before heading back to the UK.[14] He would return in 1973.

Work at the Horticultural Research Area involved the study of new precision seed-drilling equipment and associated pelleted seed ('pelleted' seed was coated with clay to produce a uniform shape and size for ease of use in a mechanical seeder) and its embedded pesticides, graded seed, increased plant densities, and the impact of herbicides, nutrients and irrigation.[15] In 1968 Bob had a visit from Joe Ward of Stanhay, which led to a connection with Germaine's in California, a pioneer in pelleting seed and therefore highly relevant to Bob's research.[16] Bob was curious to visit California to see for himself how the industrialisation of horticulture had transformed the sector in recent years and managed to secure funding from Monsanto for a study tour of intensive vegetable production sites in North America and Europe.[17] The tour was planned for 1969.[18]

Monsanto also granted Bob funding for research into herbicides and other agricultural chemicals, and noted its appreciation for 'the work and help you have given Monsanto in evaluating our herbicides and their performance under New Zealand conditions'.[19] This gave a fillip to Bob's PhD research, which he had begun – with tomatoes – in 1968.[20] His research had an inauspicious start when a massive hailstorm practically wiped out his trial.[21]

In March 1969 in a letter to Stanhay in the UK, Bob commented that his direct-sown and pelleted cabbages and Brussels sprouts were 'impressive'; 'the most spectacular advance has been in the precision direct sowing of pelleted tomatoes for processing'.[22] In the same month he published an article in the *New Zealand Journal of Agriculture* in which he revealed the progress of his trials. His comparisons between direct-sown and transplanted cabbages, onions and tomatoes left no doubt about which system had the most promise. 'The future of New Zealand as an exporter of fresh and processed vegetables', he wrote,

> will depend upon the ability of the country to produce a cheap, high-quality article that can compete favourably on world markets.
>
> To do this, growers must exploit new techniques in order to reduce costs of production and increase yields per acre. Investigations in this applied field must begin now.
>
> The exploitation of greater plant densities requires that each of the following advances be utilised to the full: precision drilling for accurate seed placement;

seed grading for uniformity in regular seeds; pelleting for precision drilling of irregular seeds; and pesticide and fertiliser incorporation.[23]

Bob's 1969 trip took him to California's Salinas Valley, Davis and Berkeley, where he witnessed a holistic vision that brought together new crop development, precision-growing techniques and mechanical harvesting. The horticulture sector – and the environment – had been entirely transformed by mechanisation. Bob recalled, 'Their fields of tomatoes stretched as far as the eye could see.' The mechanisation of tomato harvesting was, he thought, 'one of the most classically brilliant pieces of work'. He was impressed by 'the vision of the scientists, and … the skill of the Californian university education system which links theory and practice so intimately together that they were able to change the whole industry in ten years.'[24] Between 1960 and 1969, production in the Californian tomato processing industry had jumped from 2,249,000 tonnes to 3,372,600 tonnes. The acreage of tomato production had grown, too, from 52,610 hectares to 62,321 hectares. Total income from processing tomatoes had leaped by 114.7 percent, from US$52,627,000 to US$112,982,000 (by the end of the 1970s this had grown to US$428,625,000). This growth was quite out of step with the US tomato processing industry on the whole: acreage actually dropped over the decade by 3.9%, tonnes harvested increased by just 20.9%, and total farm value from tomatoes increased by 60.8%. In other words, what Bob saw in California alone in 1969 represented 66.4% of the value derived from the entire US tomato processing industry.[25]

However, the social impacts of this change to mechanisation were massive and detrimental. Californian rural studies commentator Ildi Carlisle-Cummins has described how many field workers spent their days on the back of a harvester and experienced motion sickness from looking down at the tomatoes as they were gathered along the two-kilometre rows. Mexican farm labourers – provided through the Bracero Program, a short-term labour accord between the US and Mexico – were displaced. Eighty-nine percent of tomato growers in the region, unable to afford the new machines and the land to make them financially viable, had actually gone out of business, and some 32,000 farm workers had lost their jobs during the transition.[26]

Carlisle-Cummins wrote that the new machinery, provided through the University of California, worked against the public good requirements of the university's funding arrangements. Many of those adversely affected by the changes formed the California Agrarian Action Project, which in 1979 sued the university on the grounds that the mechanical harvester had violated the 1887 Hatch Act, 'which provided federal dollars for agricultural research intended to support family farming'.[27] This case ultimately had important consequences in stimulating a revival of family farming; and the university established the Small Farm Centre in 1979 in an effort to repair its public image, a move that prefigured some of Bob's later work in New Zealand.

Impressive though the revolution was, it clashed with a wholesome agrarian vision that Bob (perhaps unconsciously) had been harbouring, and which he had witnessed in the farming estates around Wiltshire in his boyhood. Despite the impressive scale of the research and extension programme, he admitted he was 'disillusioned'. The work at Davis, California, was 'less stimulating than expected'. He commented to his parents that he felt his own work would stack up well in comparison, however.[28] In his formal report on the visit, he stated that the growing techniques he had witnessed 'lacked the precision that we strive for here at Lincoln, yet the mechanisation of harvest, packaging and transport were magnificent in their efficiency … I felt very strongly here more than elsewhere in the United States that a very good topic for investigation would be the social implications of this type of progress.'[29]

Bob's visit to the US coincided with an upsurge in anti-Vietnam War protests. These were particularly apparent in the university environment, where the students seemed increasingly radicalised. 'Certainly the situation is very serious,' he wrote to his parents in June 1969: 'The streets look like something out of the Eastern Bloc countries, grim-faced troops fully armed … preventing access to certain areas and many shops with boarded up windows.' There was something exciting for 30-year-old Bob in this state of open rebellion:

> The people in the streets are incredible, the smoking of pot is very common, probably equal to smoking and drinking, and the generation gap is massive. I've found the students vital and very aware of their situation, full of discussion, questioning the whole establishment and not happy with the answers. Despite all this there is terrific hope, as more and more people join what can only be classed as a great socialistic resurgence it is becoming harder for the National Guard to keep order and their violence wins ever more supporters for the left and the right goes further right towards suppression and violence. The last few weeks have seen death and destruction here seen usually only in under-developed countries, over 10,000 people marched on the Capitol in Sacramento last week and on Friday 30,000 marched in Berkeley and the marches were peaceful.
>
> … It is stimulating and heartening to find oneself in the company … [of] people who think as I do … Socialistic thought in America is probably a majority with the young people, and as big companies take over the running of the country to a greater and greater extent I can see no easy answer for the young people of America … Boy O Boy is NZ going to be quiet after all this.[30]

As well as squeezing in an enjoyable visit to his cousins, Uncle Cyril's children, Bob travelled to Michigan in the Midwest to see the research with cucurbits and other vegetables taking place there, then to Britain where he visited his parents.[31] Bill and Madge found Bob changed. His friend Ian Thompson visited them sometime later and wrote to Bob that they had found 'your own particular dynamism hard to comprehend when you were there last, reciting your ambition to throw parties et. al.'[32]

Britain was also the home of the Stanhay Drill and, without seeking permission from his department, Bob purchased one. To his surprise, he was allowed to keep it. While on a visit to another horticultural institute, Pershore College in Worcestershire, he met with Bob Douglas, who was now studying for a Royal Horticultural Society diploma with a specialisation in vegetables.[33]

Monsanto's manager of development in Europe had arranged a schedule for a proposed stopover in Belgium and the Low Countries, but instead, at the invitation of an academic there who had been at Lincoln for a year, Bob chose to visit Czechoslovakia to explore the fruit industry there. He visited a collective farm, 'which was most impressive', and was heartened to see 'their achievements both socially and in the field of horticulture': 'Collectivisation of the farms has achieved very good progress and the paddocks of onions, cucumbers and carrots looked more reminiscent of California than Europe.'[34] He wrote afterwards that this leg of the trip 'was really quite a highlight'.

Bob had originally planned to visit research stations in Soviet Russia but was strongly advised not to by Monsanto. Once he had returned home, he wrote to Monsanto to describe his exciting time behind the Iron Curtain, 'dodging Russian soldiers and trucks and trying not to look Western while keeping my camera well out of sight'.[35]

•

In New Zealand Bob was developing a name for himself in the local horticultural community. In 1968 he had published a bulletin on machine harvesting of onions and a booklet titled 'Extensive Vegetable Production'.[36] In 1970 he published a long article in the *New Zealand Journal of Agriculture*, giving preliminary results of his work on the use of pelleted seed in onions. The work seemed extremely promising, and he even dared to hope that 'by perfection of these techniques it will be possible to produce onion crops of such a constant size that grading will become unnecessary'.[37] One of the particular advantages, he said, 'is the incorporation of thiram and mercurous chloride into pellets to control smut and white rot in infected areas'.[38]

Bob trialled a range of tomato varieties in quick succession. In 1968 he had grown varieties of 'high solids' tomatoes, which were better able to withstand the harvesting process due to their firmer flesh – as he said, 'the original harvesters were pretty rough'.[39] Varieties included old ones like Fireball.[40] Now he was also trialling some new mechanical harvester varieties. In August 1971 he recorded germination rates for the variety 'Mech. Harv. 4', and on 11 October 1971, the farmers he was working with in Waipara, north of Christchurch, direct-drilled Fireball, 4 and 4P.[41] A few days later Bob applied the pre-emergent herbicides Dymid and Balan along with Sencor and Treflan, then drilled seed of Fireball and 'Mech. Harv. V.F. 145-21-4P'. This last became a firm (and well-known) favourite of Bob's, which he kept going for decades.[42]

The tomato trials in Waipara were associated with the movement of the grape industry to Canterbury. Bob recalled how the lecturer in fruit at Lincoln, David Jackson, wanted to grow grapes: 'Everybody pooh-poohed that of course ... I thought, if you can grow grapes, you can grow tomatoes outside.' He discovered that the Forest Research Centre at Waipara had kept weather records for many years that showed Waipara was warmer than Lincoln. Bob's PhD research on onions and tomatoes was mostly funded by the Canterbury-based vegetable processing company, Goldcrum Hayward. 'I discovered the Waipara Basin, and its microclimatic advantages essentially for tomato production, because Goldcrum Hayward wanted to extend the season.'[43]

Bob particularly wanted to learn how tomatoes could be direct-drilled successfully, rather than transplanted as seedlings, which was costly. His work was aided by research conducted by Ken Young, who was exploring the use of windmills in frost protection. Young's work identified the likely first date of a killing frost in spring, which led to Bob's realisation that it was possible to drill a tomato crop on 17 October. If a late frost occurred in November, the plants would still be close enough to the ground that they would be enveloped in the microclimate of the soil–air interaction. 'And we never lost a crop.'[44]

Local industry was already adapting to the changes, and much of the pelleted seed Bob used was produced in Sockburn, Christchurch. From 1971 his field notes refer to the availability of pelleted seed for gherkins, asparagus and celery.[45] Although the research in this quickly expanding field was exciting, Bob told his parents in July 1971 that he was 'not particularly happy' and was 'thinking of "opting out"' of his work at Lincoln.[46] His friend Ian Thompson picked up on this unhappiness: 'Your last letter was pretty grim ... Things must be grim to have *you* write so distressing a letter.'[47] Towards the end of 1971, Ian reiterated the sentiment:

There is something hard to define about many of the things you write. They are
statements of a person who sees himself trapped in time, the 'we will do our own
thing in our own time' and 'life goes on, I try and make it mean something' type
statements are all about you in a particular time. I wonder what causes a man to
write those sorts of things: the conflict between who he is, where he is and what
he is; growing old and not knowing nor recognising what is slipping between
his fingers but knowing that it is irrevocable; looking back on golden, but lost,
summers.[48]

There is no doubt that Bob was frustrated. He was becoming argumentative. 'I seem to remember you … wading into people boots and all and stamping on them like a crazed zealot,' Ian wrote to Bob in 1972. Ian recalled a dinner party they had both attended: 'You almost busted that up and insulted her friends, but left smugly triumphant.'[49]

It is likely this clash was related to politics, and more particularly to the Vietnam War protests, in which Bob had become very involved. On 18 June 1971 he took part in the first Canterbury University Academic Mobilisation meeting, and he helped organise an anti-war mobilisation at the Ilam campus of the University of Canterbury on behalf of the Concerned Academics Committee.[50] He became convenor of the Canterbury University Academic Staff Mobilisation, a role he took seriously judging by the number of letters he wrote to members of parliament and the work he put into helping organise events such as teach-ins and marches.[51]

Bob's political vision had become more acute in 1971. His earlier comments about the state of the world and the particular contexts he encountered (such as his discussions with the de Waals about Cuba and South Africa, and what he had seen for himself on Californian and Czechoslovakian farms) were reactions and observations. Now he was actively engaged in the struggle for particular political outcomes. His correspondence on this topic to Keith Buchanan of Victoria University seemed upbeat: 'We are getting somewhere. At Lincoln … it was, without exaggeration, dangerous to distribute literature a few years ago. Now we have a thriving socialist action group, not only among the students, but also among the staff.'[52]

He spelled out his political position in a letter to the Labour Party's Norman Kirk who, at the time, was Leader of the Opposition:

I believe there are many people in New Zealand who wish to build a Socialist
country through democratic means during the coming election. Some, thankfully,
have not yet compromised their socialism to mediocrity and have dared to
criticise Labour as it stands today … Many of us have worked with Socialist
Action during the long years of confrontation over National's support of the

inhumane annihilation in Indo China ... Socialist action within the Party can be useful, outside the Party, Labour could find it a very destructive force.[53]

He signed himself 'R.A. Crowder, lecturer and socialist'.

Madge felt he was going too far. Her letters usually consisted of remarks on the weather and the garden, and updates on David (who by now had a one-year-old son, Simon) and other family and friends, but in August 1971 she wrote, '... thought that from your letters of the last few months that you were not your usual self and not really "with us". I don't think you are very happy at the moment.'[54] She was concerned that she was losing her son to the radical agenda he had been exposed to, and so excited by, in California:

> Both Father & I felt very worried and sad that you seem to be jeopardising your career and becoming at loggerheads with the College authorities by being too actively connected with this Left Wing assoc. It must be very worrying for the Masters who are responsible for training their students to be good Agriculturists to have one of their staff distracting them and instead of them studying have them demonstrating and causing obstruction in Christchurch. You should see their side of it Robert and act in a responsible way even if the motives are good, and not expect everyone to have the same ideas. Everyone in authority must conform in certain things or things get chaotic.[55]

Madge was not convinced that political action would make a difference:

> We all know the world is in a mess but Politics, Right and Left and organised religion have in the past caused most of the hate and injustice to man and they are still one of the worst evils. You are wearing yourself out and getting obsessed with ideals which of course all sane people would like to see come about, but unless man's character is changed throughout the World it is just impossible ... I am sure you can contribute far more to the wellbeing of the World by sending out your students well trained and happy.[56]

Madge noted that Bob's back problems had prevented him from tramping, 'but don't let that turn you into an "Angry Young Man". We have had plenty of those for years, but generally they find there is more to life than that and get married and have a family and find the World is not all Evil.'

Finally, she invoked God:

> I don't know whether you still believe in God and the Bible, Robert, but if you do you should remember that it is still God's World and he will prevent it from destruction when the time is ripe. The Associations who study the Bible have always said that according to prophecy the world would come to the stage it is fast approaching now, and when you see things happening now, which I heard about when I was quite young, I think maybe the final struggle between the Nations

won't be far off … so maybe you will see the beginnings of a perfect World, with peace and justice but not under present man's rule.[57]

Bill also took to the pen:

I feel much sympathy with your aspirations for the underdog, am genuinely horrified at the seeming brutality of you know who in Vietnam but also at the same sort of thing that has happened in Poland and other countries on that side of the Iron Curtain, and probably are still happening.

I am certain that Capitalism is wrong for many people and that it is slowly going to a point where it will blow itself to pieces, but at the same time I hope whatever takes its place will not entail the bloodshed and horror that major changes in the past have always thrown up.[58]

Bob continued to mix with disaffected young men. One of these, John Dickinson, wrote to Bob from Brisbane early in 1972 to wish him a happy Christmas and prosperous New Year, adding acerbically, 'what a capitalistic word'.[59] John joined the Australia Party (an anti-Vietnam War party) in 1972 in the hope that 'with their preferences the A.L.P. [Australian Labour Party] should be assured of winning the next Federal election'.[60] Unlike Bob's parents, John was pleased to hear that his friend's anti-Vietnam demonstration was a success.[61] In 1973 he wrote that he was now very involved with the Australia Party: 'It still is, thank heavens, very idealistic even though (you would not like it) not very socialistic – very strong on conservation & liberal moral issues e.g. legalized prostitution, homosexuality, liberal population control, abortion laws etc, more control of polluters … Also the AP is very peace orientated.'[62]

Howard Gill, who began lecturing in Lincoln's Horticulture Department in the early seventies, shared many of Bob's values and was an important influence.[63] Howard was a 'radical', and his later work included providing guidance for trade unionists. Raised in Northern England, he was the son of a railway worker, and his view of the world seemed, to Bob Douglas at least, to be very 'us and them': the workers and the bosses.[64] Bob Crowder and Howard had regular political conversations, and the two of them, along with Howard's wife Titi (who was a sociologist), took part in protests against the war in Vietnam.

Bob was becoming increasingly uncomfortable with the direction his work was taking him. His PhD research was essentially about enhancing efficiency and monetary gain at the expense of social wellbeing: 'It worked against my philosophy.'[65] Howard Gill's research, which was largely around setting piece-work rates for pickers, showed that, managed properly, hand-picking could be as profitable as machine harvesting with the added bonus of more money going into the local economy.[66]

'Basically,' Bob said, 'I agreed with him.'⁶⁷ He avoided making public statements on such things, and anyway, his research was more about the improved methods of growing a crop rather than the harvesting, per se. His 1972 *New Zealand Journal of Agriculture* article on the results of his asparagus trials argued that there was 'no doubt that direct-drilled asparagus … offer[s] an economical and rapid method for the establishment of commercial asparagus plantations, especially when processing quality is the main object … Such developments will entail the greater use of fertiliser, irrigation and probably the complete reorganisation of handling methods for this crop.'⁶⁸ But in April 1972, about the time when his political interests seemed to be peaking, his field notes began to mention the harvesting rates of his crops (weights, dollar values and hours involved). They also contain an oblique reference to National Party Prime Minister Keith Holyoake, the blackmail of Labour Party members in marginal seats, and United States military matters.⁶⁹

While still residing on campus, Bob continued to organise social events. He introduced ballroom dancing, but his most famous social contribution was a monthly disco in Memorial Hall, which ran from 1969 to 1974. The Village Gate performed after 10pm once their gig in town had finished.⁷⁰ There was no bar at these discos; to keep drinking under control, Bob organised the young men from his classes to provide a full steward service. Punters ordered drinks and snacks from a list and the stewards brought out the orders.⁷¹ He was the first person at Lincoln allowed to run an event with alcohol.

In 1972 the drama club lamented that it had degenerated into near obscurity, 'which leaves the campus community with what could be described as S.F.A. – and we all know how much that is'. The two plays they had managed to put on during the year were both presented 'in conjunction with one of Bob Crowder's discotheques in the Memorial Hall'.⁷²

Bob ran a 'pagan feast', with Leon Langley of The Landing on catering: 'It was all whole chickens … there was a dais where the feast was held, there were rushes on the floor, and candles, a great big pig's head on a stand and … dancing out on the floor, again with wine service with stewards.' The evening started dramatically with Wagner's *Ride of the Valkyries*. Suddenly the curtains opened, revealing a woman dressed 'Brunhilda-style' in a helmet with horns, sword aloft, in front of the candle-lit tables. 'People ate with their bare fingers.' He also organised catering for the graduation ball, a significant event in the college calendar.

From the doldrums of 1971, it seemed that things were improving. Madge was glad to hear that his tomatoes 'had been a great success so far, & that everything is going well for you. You work very hard and it's good to know you are getting results and we are very proud of you and hope you get your PhD in the not too

Asparagus trials, 1970s.

distant future.'[73] Her letters spoke largely of her distress about the rising cost of living in Britain, on-going strikes and electricity cuts, the troubles in Ireland and, above all, her own ageing and inability to keep on top of housework and gardening. The uncles were dying, too:

> We shall miss Uncle Jack especially at Sunday dinnertime, he has been coming over for so many years, he was the last family we had, now we are really alone with no-one belonging to us in Gt. Britain. It has all happened so quickly since Nov last year. First Uncle Bert, then Percy now Uncle Jack. We shall soon all be gone.[74]

She dreaded 'starting the gardening again I am so full of aches and pains and without much energy I don't know if I shall be able to cope … Trouble is I'm getting an old woman, 65 now, and I don't much like it.'[75] Bill was planning to retire. He wrote early in 1973 that although due for retirement in April, he would stay on until the end of the year, 'by which time I hope to have adjusted myself mentally to being in the junk heap. I will then take on a part-time job doing the garden regularly and helping your mother with the housework.'[76]

In 1969 Bob moved into the Miles Warren flats at 64 Carlton Mill Road. The flats enjoyed a spectacular aspect onto a large lawn that swept down to the Ōtākaro Avon River and the bluebell glades on the other side. In return for mowing the lawns and keeping the gardens tidy, Bob received a discounted rent. A friend wrote to Bob in March 1970, agreeing that 'the move was overdue … now you can start to live a little, and by your letter you have been'. He was glad that Bob seemed 'happy at last', that he could now 'let your own personality out', and how 'unbelievable' it was, 'after all this time, to suddenly find yourself one of the fellows' – in other words, gay.[77]

Bob confided that he had met a married man at one of his dinner parties. 'I always like the married ones,' his friend replied, 'as they are no problem, never get too keen as they have the family to think about … You can learn from him, as he must know the ropes.'

> The only thing I forgot to teach you were the different ways one has sex, you better ask your married friend as there are so many different ways. Just don't go mad, take it easy … if you go from one place to another, people talk. Play it cool & be careful, take a bird out now & again … play it square until the lights go out, then the sky's the limit.[78]

Bob's aspirations to throw parties at his own place, which had so startled Madge and Bill, were now realised. The lawn at Carlton Mill Road saw many fabulous gatherings, and his flat became famous for the Boxing Day parties he held for 'misfits', sometimes with 80 or 90 people on the large lawn that sloped to the river.

Bob socialising at his Carlton Mill Road flat, 1970s.

One of the misfits was Kevin Bussell, who ran the Victoria Coffee Lounge on Montreal Street, which Bob frequented.

With Leon and Kevin in the mix there was often a culinary angle to these events. Globe artichokes featured, a vegetable of interest to Bob and to Leon, who was of Portuguese descent. Bob's interest in globe artichokes stemmed from his 1969 trip to California, where he visited the Salinas Valley. 'There I met an old Italian who gave me a little packet of globe artichoke seeds, and he said I needed to take these back to New Zealand with me and grow them.' Every seed was of a different variety. After years of tastings, 'Leon selected the one that he thought was the best, and named it Crowder's Delight.'[79] In 1975 Bob propagated 62 plants of Crowder's Delight. Leon began to include globe artichokes on the menu at his restaurant, where they became popular. The Dux de Lux, a prominent restaurant and bar in the Christchurch Arts Centre, and the restaurant at the Sign of the Takahe were among other takers.[80]

Bob later recalled his time at Carlton Mill Road as the best seven years of his life.[81] The parties and the view over the river were high points, but so too was the communal nature of the Miles Warren flats. In 1992 he looked back on this as a kind of model for good living: 'We had one washing machine for four flats. Our social contact revolved around the wash tub.'[82] He had experienced communitarian living and loved it.

Although there were plenty of gay friends around Bob, there was still a need for discretion. He wrote to Ian Thompson in 1971 saying that Kevin, who was married, 'drops in occasionally for a little playing'.[83] 'I wonder how much of a lie that marriage is,' Ian remarked.[84]

Bob confided to Ian about his love life, prompting Ian to write: 'It's reassuring to know that you are well … and reassuring to also know that the "people" you meet, even Australian boys in strawberry patches, only "meet a need of a moment" to quote you. That saves me thinking, like Ulysses, I might one day be slaying suitors.'[85]

On one trip to Auckland Bob met Peter Kupu, a Tongan man who eventually seduced him and came to Christchurch to visit. In search of adventure after his work of several years, Bob visited Peter in Tonga over the Christmas break in 1972. He travelled via Fiji, where an air strike kept him for a few days.[86] The holiday was 'very strange & challenging', he wrote to his parents, but was 'just what I needed, and being insecure does one a world of good'. In Fiji he had found himself among the locals at the cheap bar in the Hotel Suva and was 'soon being taken around the place'. Once in Tonga he attended two traditional village feasts and enjoyed the pigs, fish, vegetables, pineapples, coconuts and melons.[87] He was appalled, however, by the poverty and the hint of corruption.

He stayed with Peter's family for three weeks, immersed in a Tongan community, and commented,

> I must say their way of life is proof of the benefit of simplicity. Never did I hear a voice raised in anger against children, never did I find Tongan children obtrusive or a nuisance. They all practise socialism in the name of Christianity and would share their last taro leaf with you … My strongest impression of Tonga will always be the tremendous family loyalty and unity and the very extensive living witness to Christianity which I've never seen before and also the wonderful children who are literally so innocent, so carefree and so incredibly happy with absolutely nothing.[88]

He was struck by the privilege and power of the nobility and the negative effect it was having on the local people and environment. He quoted someone from the American Peace Corps who told him, 'Here we are trying to create that which we are escaping from in America.' Bob had some 'really vicious arguments' with Peter about Western influence:

> What the future will be is easy I think to see and made my holiday in some respects rather sad. One asks oneself can I help at all, how can you make these people see what the inevitable consequence will be and ways and means of altering the pattern.

> We are bastards is the only conclusion I can come to and tourism is the worst bastard of all. It creates no greater understanding of peoples only spreads greed into the minds of poor people as they see the apparent glitter of a superior culture.
>
> This is why I support Communism so much because it offers an alternative to the way they are going, and Tonga with its socialistic Christian ideals could so easily set an example for other small countries to follow – but I doubt whether it will.[89]

(Bob later repudiated the idea that he was a communist, clarifying that he really meant he supported socialism.[90])

While on holiday he turned 33. In a surprise twist, he 'adopted' two 11-year-old boys financially, with the intention of putting them through high school and, 'if necessary', college or university. The boys were Peter's nephews.

> Hope is that by education they might see the peril their country is in and do something about it, but I suppose really I did it because in the end people matter more than ideals and when you get to know these people then you want to help them even if it is to better equip them in our far from perfect society.

He promised to send Madge and Bill a photo of 'my new sons'.[91]

His reflections on American and Western influence in the Pacific amplified his thoughts on the Vietnam War, which was drawing to a close. 'Well done, N. Vietnam, determination and a just cause have destroid [sic] American arrogance for once.'[92]

Bob's relationship with Peter did not develop much further, and Peter, who in 1973 was living in Tonga, announced that he was to marry an American woman and invited Bob to be his best man. Shocked by the news, Bob declined. Peter explained that the marriage was 'a pretty good chance to get out of this … nasty society of Tonga & we can come up to see you quite often, or she may stay behind while I will visit you'. In 1974 Peter would argue that 'there is nothing whatsoever to stop our friendship', and the following year, from the United States, he wrote that it was still his dream 'to think & to talk and be with you'.[93]

But Bob had moved on: at the United Services Hotel he met Stephen Craik. 'Our eyes met across the bar,' he recalled. They ran into one another again later the same evening at a party.[94] Bob's friend John Dickinson congratulated him: 'Pleased to hear you have settled down with someone – it is necessary to have some form of stable relationship in this "topsie-turvey" world – Hope everything works out for you.'[95] It didn't last. In September Ian Thompson wrote to Bob saying, 'he [Steve] didn't know what was going on in your corner of Carlton Mill Road … Just as well, better than tearing at each other, better off as kind friends than something like embittered lovers.'[96] He went on:

I don't much like the 'Probably we are all that we have got, pathetic really' sentence. It not only smacks of a morbid feeling sorry for one's self but it's so untrue, look about you and see just how much you have got, how many folks about you are there that regard you as theirs in a very special sort of way and see if that's pathetic. At times you undersell yourself, and you are too important, have too much to give to do that. Perhaps you just need to be told that more often.[97]

•

Throughout 1972, Howard Gill's studies into handpicking perhaps helped to sharpen some sense of meaning in Bob's work. He broadened his tomato repertoire to include 'V.F.134-1-2', New Yorker, Red Top, Spring Set, Roma, Napoli and Heinz 1706 in 1972.[98] It is clear from his 1973 article on onion yields, published in the *New Zealand Journal of Agriculture*, that he was investing considerable time into his detailed studies and was respected in his field for this work.[99] Yet it seemed to Ian Thompson in 1973 that Bob was 'all weather and tomatoes and stop the world and let me off'.[100] Madge felt that Bob's letters had 'changed over the years. You were so full of your new life then and how wonderful NZ was etc.'[101] Bob reported to Madge that he wasn't getting time to do his PhD.[102] But if he was glum again, at least there was a trip to the United States coming up, which Bob imagined would be 'a hunt for beautiful people among the cucumbers' – a line of thought that Ian Thompson approved of.[103]

Officially, and 'with the support of commercial interests, impressed with the value of his work on field-scale production of tomatoes and other process crops', Bob spent a month in the United States from mid-August.[104] The focus of this trip was actually 'to witness the growing, harvesting and handling of pickling cucumbers', an unusually complex matter as Bob revealed in his report of the visit.[105] Goldcrum Hayward was the commercial interest. The company supported his work in other ways as well, notably by paying half of the airfare to bring Bob Douglas back to Lincoln in 1973 to commence work as the technician at the Horticultural Research Area.[106] Douglas acknowledged that he returned because of Bob Crowder; he considered Bob inspirational and a 'big thinker … twenty years ahead of his time'.[107]

Douglas worked at Lincoln with Bob until 1980. His main tasks included looking after Bob's research, which meant doing a lot of weather records, managing plantings and replicating trials. He was then promoted to manage the Horticultural Research Area, where he worked with other academic staff (including Howard Gill on the labour studies) and gave practical talks to Bob's students.[108] The research area expanded to include apples and nuts, but the main work on intensive precision vegetable production continued. In Douglas's view, the research at Lincoln compared favourably with that being carried out in California.

In 1974 Bob expended a huge effort on his disco in the lead-up to the Commonwealth Games, which were held in Christchurch. Lincoln College was to be one of the main residential hubs for the Games and visitors needed entertainment. Bob and Leon Langley would run the disco every night. Plans were changed at the last minute, however, and only three discoes were held. They were the last that Bob ran.

Bob's parents were increasingly stressed by the deteriorating situation in Britain and televised coverage of the Games brought Christchurch into their living room. 'Father & I both thought it all looked very nice & if things get much worse in Britain and we become refugees we shall know where to make for.'[109]

Untargeted frustrations were affecting Bob's life. In 1974 he wrote to his parents, 'I'm going through a bored crisis and need a new challenge to keep the adrenalin going.'[110] He applied for a job in an agricultural college 100 kilometres west of Brisbane but was unsuccessful. Instead, he was promoted to senior lecturer at Lincoln.[111] But after 10 years of working with chemical companies and travelling around the world, Bob had come to the conclusion that 'it was not the world I want to live in'. The thought of writing up his PhD was not appealing; in addition, poor weather had been affecting his projects.

In 1974 Bob published what would be his last article for the *New Zealand Journal of Agriculture*, outlining in detail the results of seven years of work on tomatoes. He concluded that commercial tomato growing in Canterbury was viable, but he was equivocal about machine harvesting, saying, 'We can harvest them with a machine for a price that will not be much different from hand harvesting.' It was not a ringing endorsement.[112]

His father wrote to him about the poor weather, hoping 'that it has not been too serious and that the results will not be too bad'; a week later he wrote that he was 'worried' and waiting 'anxiously' for news on this point.[113] By mid-March Bob had managed to get most of the crop in and it seemed, to Madge at least, that he was 'seeing results' after all his hard work. He was even featured in the local newspaper, bottling tomatoes, and made a television appearance as well.[114] But Bob was struggling.

He voiced his concerns to Mac Morrison, who suggested he look for a job elsewhere or go on sabbatical. Mac was interested in a new teaching style being developed in Bath, in Britain, the so-called 'sandwich degree', which combined practical work in industry with formal class content in a structured manner. He suggested Bob go there to check it out. Spending time in Bath seemed like a good option to Bob, and his parents would no doubt be happy to have him close at hand

Bob bottling tomatoes in his Carlton Mill Road kitchen. (Photo courtesy of the Christchurch Star)

for a year. He applied for 'refresher leave', which was granted in May 1975, and received a formal offer in July from the University of Bath for the role of temporary lecturer in the School of Biological Sciences.[115] As predicted, his parents were delighted. 'How exciting to hear from you ... and to know that after 6 years we are again going to see you,' Madge effused.[116]

This news may have helped calm Madge somewhat; her agitation over the sudden rise in prices was growing more fervent, and she poured out her political views to her son:

> If we have your left-wing Gods in power for long in the Labour Party especially that horrible Benn the country will be ruined completely ... and still the stupid workers under the influence of these extremists go on strike for more money, that's all they think of, greed for money even if they will be pricing their products out of the market, and causing most of their more decent thinking fellow workers to lose their jobs eventually. Unless we get a strong government soon who will put the country before politics and stand up to the Unions I think we have had it & we shall all be on the bread line, the ordinary middle class & working class, but it won't be the rich.[117]

And so, in 1975, Bob returned to Britain. The move heralded a profound shift in his professional trajectory.

Bath City Morris, Bath, 1975.

4: Organic conversion, 1975–83

While on sabbatical at the University of Bath in 1975, Bob made a conscious decision not to undertake research. Most of his time was taken up with work in the School of Biological Sciences. The Bachelor of Science in Horticulture at Bath combined practical placements with classroom learning – a 'sandwich' model – and Bob found this style of applied learning a refreshing challenge. As well as lecturing he got to visit many of the best vegetable producers in Britain while checking in with students on placement.

His search for suitable accommodation in Bath led him not to 'regency splendour in the better terraces of town' but to a leaky flat down by the station, 'over a cafe by the side of a fish and chip shop with three pubs within spitting distance'.[1]

While in his office one day, Bob heard the jingle of bells outside – and discovered the Bath City Morris dancers. He had intended to take up Morris dancing in Britain and in Bath it became an important part of his life. He joined the flourishing folk music scene, too, and in his notebook described the Hat and Feather, a folk music club on Walcot Street. Bob surmised that the popularity of 'folk' was stimulated by the energy crisis and a growing interest in sustainability and self-sufficiency. No surprise then that just down the road from the Hat and Feather was the Bath Arts Workshop, which was connected to the Bath Community Design Workshop, later known as Comtek. Comtek was concerned in particular with 'energy conservation and the interdependence between each person and the local community'.[2] Bob scribbled the address in his notebook: such thinking interested him.[3]

This was also the year in which Bob came to terms with his homosexuality.[4] He met Chris Weeks at a gay pub, the Theatre Royal Green Room, and the two became partners for the year. He also made another gay friend, Martin Stokes, who was working on tissue culture in one of the laboratories that the students visited.

Bob spent time with his parents in Devizes. The garden he'd created there was 'a great success'.[5] When he returned to New Zealand later in the year, his father Bill expressed his gratitude for this time: 'It was a momentous year, and one we shall be happy to repeat … we loved every minute.'[6]

•

Bob made his way home via Philadelphia and Boston, where he visited horticultural sites and spent a week at the University of California with Bill Sims, who was also engaged in tomato research. If he was not sure of his future before the trip to England, upon his return in August 1976 Bob was clear: 'There is no doubt that England is my home, I settled in so well & enjoyed it all so much.' He had 'enjoyed having a mum & dad again and dropping in for dinner, teas etc.'[7] But any thought that he might return to Britain to live was put to rest: 'I shall set to [and] buy my little property over here in NZ because in the final analysis I am better off here and the restricted opportunities at Bath would have soon led to a narrow existence for me.'[8]

His research work at Lincoln was proving problematic. The weather was not behaving and the season in his absence had been disastrous: 'Poor old Bob Douglas had to cope with that.' The season following his return was also poor. Bob's major source of funding, Goldcrum Hayward, left the district, and these factors together spelled the end of his PhD research. He never returned to it. He did continue with his vegetable trials, however, and in November 1976 noted the post-emergence herbicides used on the onions: Sencor, Tolril, Linuron, Ingran 50, Tribunil. The tomatoes received Ethrel in 1977.[9]

On the other hand, Morris dancing thrived. Once popular in New Zealand, Morris dancing had died out in Christchurch in the 1930s and in the rest of the country by the 1940s.[10] On his return to Christchurch, Bob lost little time in setting up a Morris dancing side (the term used to refer to a team of Morris dancers). In October 1976 he placed an advertisement in the newspaper and received a call from Ian Joice, who expressed interest in joining. 'From that quiet start the group gradually came together using Carlton Mill Road as the first meeting ground.' They later moved to a room at the Arts Centre.[11]

Bob became the squire, teaching and leading a side that included women – a point that quickly became political. His approach was typically practical: 'Originally it was intended to develop the Morris tradition of dancing based on male dancers only but this proved impossible due to the scarcity of men wishing to dance ... either one encouraged everyone or [one] forgot the whole concept of Morris dancing in Christchurch.'[12]

Bob spearheaded the Morris renaissance in Christchurch and appeared in full Bath City Morris dancing costume in the *Press* in 1976. 'If you enjoy vigorous dancing and are superstitious enough to believe it could bring fertility to your vegetable garden then Morris dancing could be your ideal hobby,' the article read.[13] Judith McArthur wrote, '[The dances] have always been associated with the permanence of the countryside, the simplicity of rural life, and, most of all,

Erewhon Morris compete on Opportunity Knocks, *1977. From left: Ian Joice, Elwin Jamison, Fleur Lester and Bob.*

a cheap way of having fun.'[14] In public, Bob highlighted the benefits of the dance form: 'The attraction of Morris dancing is its basic traditional beginnings, it is good fun, colourful and above all, very friendly. A terrific rapport is established between public, dancers and musicians and displays are always informal and relaxed.'[15]

By early 1977 the side was relatively stable and needed a name and costume. A visiting Morrisman from Perth suggested 'Erewhon Morris'; the owner of Erewhon

Station was duly approached for permission to use the name and invited to become the honorary president. He not only accepted, but donated to the club 'its fine set of Merino horns and fleece as the Hobby' – a kind of mascot that travelled with the group.[16]

Erewhon Morris was ready for its first public performance at an Easter folk festival at Motukarara Race Course, and in June they appeared on national television on the show *Opportunity Knocks*. They attended the Whare Flat Folk Festival in Dunedin and toured Central Otago in the 1977–78 summer holidays, and for several years this became something of an annual tradition. In June 1978 they danced in Wellington and featured in the *Dominion*; they also danced for Lincoln's centenary. In November 1979 Bob and the Morris side appeared in full costume on the front page of the *Press* as part of Haywrights' fiftieth anniversary pageant. Shortly afterwards they were featured in the *Otago Daily Times*, sharing the front page with the news that the Soviet Union had invaded Afghanistan.[17] During 1979 the side danced publicly on at least 29 occasions, and by the late 1970s Morris sides had developed in Auckland, Wellington and elsewhere.[18]

Bob's prominence in the scene probably helps to explain why the later Green movement in New Zealand was often associated with Morris dancing – at least in the popular press. Morris dancing evoked the vision of idyllic agrarian life that Bob had nurtured for so long, and which clashed so obviously with his research work at Lincoln.

Chris Weeks, Bob's partner from Bath, arrived at the Carlton Mill flat on New Year's Eve 1976 and settled in. Madge enquired, 'What's he going to do about a job when he arrives or is he taking a holiday? What's happening to all his dogs and animals, is he taking them as well?' She would be left wondering. She grumbled, 'Everybody seems to be able to afford to visit New Zealand except the Crowder family. Well at least we can afford it, but I can't see it ever being possible – we are getting too old.'[19] If Bill and Madge had their suspicions about Bob's relationship with Chris, they didn't let on. Madge continued to press Bob to settle down and find a wife. He would then 'have some-one to work for and companionship and roots and a purpose in life … there is a lot to be said for family life & one gets a lot of satisfaction even if one gets ups and downs.'[20]

•

The withdrawal of Goldcrum Hayward and the subsequent collapse of Bob's research trials were perhaps a blessing in disguise. Given his increasing distaste for industrialised vegetable production, displayed so publicly in his association with the folk scene, his research topic was becoming untenable.

The lessons from Bath about the benefits of applied learning gave Bob a new and invigorating project: to establish a similar 'sandwich' degree at Lincoln. He had witnessed first-hand what a difference a lengthy, structured and assessed component of applied work could make to student competence and hence employability. But there was pushback from some quarters to the proposal to introduce this structure – Bob called it 'stiff opposition' – which he thought stemmed from 'professional jealousies'.[21] Some staff felt that students would spend too long out of the classroom and the programme would therefore lack academic rigour. So difficult was this challenge from colleagues that Head of Horticulture Mac Morrison resigned from his position and eventually left the department in 1979.[22] Bob helped to establish the new programme nonetheless, and the first intake of students was scheduled for 1978.

When Bob returned to teaching in the second half of 1976, a number of students began to express dissatisfaction with Lincoln's lack of focus on environmental issues.[23] Ecologist Haikai Tane recalled that Lincoln students complained to him about the lack of such content as early as 1974.

Haikai Tane (formerly Ron Mathieson) had graduated with a BA Hons and LLB from Canberra (1968) and held an MSc (Resource Ecology and Regional Planning) from the University of British Columbia (1972). Since coming to New Zealand in 1973 he had been cycling around the country, and in early 1974 he worked for a few months at Dunedin Botanic Garden 'to learn the local flora'. There he met Tim Porteous and Angela Meechin, who were gaining practical experience before tackling a diploma in horticulture at Lincoln: 'They were keen to learn about the ecological farming and garden systems I studied in Canada ... and asked me how to get organics and ecology into their horticulture course ... I advised them to approach Bob with their request.'[24]

In 1975 Tim and Angela moved north to study amenity horticulture at Lincoln. In Christchurch they heard about a house in Springston called Japonica Inn. Tim recalled that a number of people lived there – 'landscape architects ... they were stylish, they were practising organic horticulture ... It was a sort of archetypal 1970s alternative lifestyle place really, it was wonderful, very colourful.'[25] At the centre of this community were Brian and Cathy Tetu; Haikai had worked with Brian on a garden project in Ottawa where they followed 'traditional Quebecois organic methods'.[26] In Springston, Brian had set about establishing organic gardens complemented by craft studios. 'By 1975,' Haikai recalled, 'Japonica Inn was fully organic, into waste recycling, organic composting, rural crafts and organic gardening. Sometimes hordes of visitors arrived at weekends to engage or just see what was happening.'[27] Students living at Japonica Inn now formed the nucleus of the push for curriculum change that Bob was encountering.

Among the visitors to Japonica Inn was Tim Maples, who was not yet a student but would in time become one of the most important influences at this point in Bob's career. Tim was on a spiritual journey that was by no means mainstream. He would have started at Lincoln earlier had he not been inspired, through his yoga studies, to travel to India and live in an ashram. Before he left in 1975, his yoga instructor gifted him one of J.R. Rodale's books on organic gardening, which Tim read assiduously. In the ashram, Swami Gitananda allowed Tim and a couple of others to build Indore-style compost heaps as described by Rodale and developed by Sir Albert Howard, and to grow vegetables for the community. As Tim recalled, 'There were vast amounts of manure and vegetation and warm weather, and I came back really inspired.'[28]

In 1977 he joined the Canterbury branch of the Soil Association and was elected onto the committee the following month. He read widely about organics, including Howard's *Agricultural Testament* and Eve Balfour's *The Living Soil*.[29] He worked in a commercial market garden as well and so had solid experience and knowledge by the time he enrolled at Lincoln in 1977 to undertake the one-year Diploma of Horticulture. At 25 he was somewhat older than his classmates (of whom there were around 60), and he exerted a not inconsiderable influence. He was part of a group of students – according to Haikai, who encouraged them – who 'let Bob know that if they were not taught organic horticulture they would enrol elsewhere'.[30]

Tim's refreshing perspective on organic food production came as something of a relief to Bob, who was 'very open' to the ideas Tim brought to the course. Partly, of course, this was because Bob was disillusioned with the kind of research he had been doing, which 'didn't really fit in with what I felt was the right way to go'. Organics was not all news to him, however, as Tim believed it to be.[31] Bob was 'very aware of rotation, and organics, and composting … I'd been practising it all my life':[32]

> As a student in the late 50s and early 60s I can never recall that the importance of maintaining soil organic matter, structure and fertility was ever questioned … But then I can never remember as a horticulture student at my university … ever learning the principles of a compost heap and certainly never seeing let alone building one.[33]

Bob was enthusiastic about cultivating young minds and allowed Tim to give a lecture to the other students on biodynamics, a controversial topic at the time.

'Sandwich degree' first year students making compost, 1978.

Throughout that year Tim behaved as a 'provocateur', challenging not only Bob but also Tom Walker, New Zealand's first professor of soil science, who expressed interest in Tim's views.

Bob's later character reference for Tim (probably written in 1980) shows the high esteem in which he held this man: Tim's 'enthusiasm for organic production inspired the establishment of the organic section in the research area at Lincoln College and Tim has been closely associated with its development and student instruction ever since on a part time basis'.[34] In 1980 he commented: 'Tim Maples … has always been a success here at Lincoln as a lecturer and a demonstrator and his grasp of the subject matter is immense.'[35]

ABOVE: *Tim Maples with Jill by the organic intensive plots, c.1978.*
RIGHT: *Bob in his Lincoln College office, c.1979.*

Like Tim, Haikai was by now also involved in the New Zealand Soil Association, and for a short time from 1977 was assistant editor of the association's magazine, *Soil & Health*.[36] In his view, organic farmers – in the Waikato at least – had 'a practical awareness of land use ecosystems. Their farming works with and cultivates natural processes. They have learnt to observe the ways of nature on their farms and over the years have cultivated vigorous life in the soils, so that balanced plant growth and healthy stock are natural consequences.' He noted that the Soil Association was 'friendly and helpful … and there was a pleasant lack of the competitive drama that inhibits mutual aid and co-operation'.[37]

The organic movement, which captured the same values he found in the Morris ethic, was making its presence felt now in Bob's work. Tim and other students were demanding organic content in the horticultural curriculum; this was discussed during lectures, and Bob was willing to comply. But there was one condition, as Bob recalled: 'I told them that if they wanted to do it, I would support them, but they would have to help … We had to do it out of sight, out of mind.'[38]

Once Tim had completed his diploma year, Bob employed him over the summer of 1977–78 to establish intensive organic plots as part of the Horticultural Research Area.[39] One of the first tasks Tim undertook was the building of compost heaps. That he managed to get around 20 of his classmates to help suggests there was a high level of interest. The work was carried out through the Horticultural Club and reported in *Lincoln College Magazine* as work done for 'Bob Crowder's Biodynamical Garden'.[40] Tim recalled, 'We were able to get the resources we needed around the college in terms of vegetable material and manure to create good composting.'[41]

While Bob did not shout about this from the rooftops, word did get around. Nineteen-year-old Geoff Barnett was interested in organics and had been reading some of the literature. In 1977 he visited the new organic demonstration garden at Lincoln, which he had heard about from a student friend. Geoff undertook the Diploma of Horticulture in 1978; in time, he would become a key player in the venture Bob was developing.

The second intake of students in the new-format 'sandwich' degree in 1979 included Jo Blakely. 'Everyone knew that Bob was the alternative … lecturer at that time,' she recalled:

> He wasn't mainstream, and he certainly wasn't accepted as mainstream. But his views were respected as representative of that side … He put up … a different way of thinking, which I think is crucial for any learner, and a young person's mind. He instilled passion in all of us on our understanding of the organics movement. He couldn't talk about the Pukekohe clay loams without talking about the damage that was being done to the biology, the whole structure of soils being destroyed

by this ongoing, continuous planting of things, and he could see it not having a future unless something was changed.[42]

Bob sowed what she called 'a great seed'. 'He challenged authority. We could see that he was the odd block in the department.' Jo perceived that what seemed to be missing in some of Bob's peers was a way to take the principles he was teaching and apply them on a large scale. 'People weren't yet able to translate that to their commercial experience.'[43] Bob eventually poured his energies into this gap.

Tim Maples forged a connection between the Horticultural Research Area and the organics movement that, in time, proved immensely consequential. The Canterbury branch of the Soil Association did not have capacity to help with the organic developments at Lincoln – they were an older set and did not have funds – but they were certainly interested. The branch included Jack Whitelaw, Jack Meechin, Mollie Chalklen and George Maslin, who were all 'really pleased to have young energy' coming on board.[44] Without a doubt, the most important connection was the way Tim brought Bob's work to Mollie's attention. Tim wrote articles for the association's magazine and paved the way for much stronger support in the coming years. Bob also began to attend Soil Association meetings, although he found them formal and not very inspiring.[45]

The national organic scene, led by the New Zealand Soil Association (established in 1941), was strong but consisted mainly of passionate backyard gardeners. The association had been based in Christchurch since 1954, so the committee Tim was involved with was essentially the national council. Although the association was active on many environmental fronts, it was a somewhat 'fringe' organisation in the 1960s and throughout the 1970s.

Internationally, the organic movement was gaining momentum, and the efforts of the various organisations were broadly in line with what was happening in New Zealand. As historian Gregory Barton observed, during the 1970s 'organic societies saw past gardens and farms, and lobbied for the protection of the entire ecosystem, which included national and also global challenges. Deforestation, wildlife conservation, and the greenhouse effect of industrial gases that led to global warming were all highlighted.'[46]

A deliberate global initiative to advance organic farming principles and practices was emerging. The International Federation of Organic Agricultural Movements (IFOAM) was established in 1972 in France to ensure that 'organic agricultural movements make themselves known and coordinate their actions'.[47] By 1975, 17 countries were represented in IFOAM.[48] The coalition was heavily Eurocentric: New Zealand and Australia were not part of it.

CLOCKWISE FROM ABOVE LEFT: *The long garden path, back garden, Ashgrove Terrace; Bob outside Tasman Hut, Mt Cook, glacial dome in background, c.1960s–70s; Graeme Thiele's strawberry production research plots, 1960s*

TOP: *The first cohort of 'sandwich degree' students laying out their individual practical demonstration of comparative growing systems, which would continue into their second year.*
BOTTOM: *Intensive organic rotational system as laid down by Tim Maples based on composting and minimal cultivation techniques.*

TOP: *View of the intensive beds and the met station, 1982. Bob ensured that learning about growing was integrated with the discipline of meteorological recording using the Stevenson screen, rain gauge and evaporimeter to authenticate record taking.*
BOTTOM: *Compost heap, 1982.*

CLOCKWISE FROM LEFT: *Front driveway at Ashgrove Terrace, with beehives in background, 1982; Bob seated between Madge and Bill, with friends in his front garden, 1982; Bob and cat in Mike Daley's amaranthus trial in the BHU agricultural rotation; Solidago (Golden Rod) covered in bees; Amaranthus and sunflowers.*

TOP: *Bob's density studies of amaranthus trials in the semi-intensive rotation, 1988.*
BOTTOM: *Amaranthus trials in the semi-intensive rotation, 1980s.*
OPPOSITE: *Bob's home garden: orchard with floriferous under storey.*

OVERLEAF: *BHU 1990–91.* (Photos courtesy of Jared White)

Comparing the ability of soils to absorb rainfall: TOP: *The BHU's semi-intensive plots have no surface water present after 100mm of overnight rain, while extensive surface water is present at the neighbouring conventional orchard (*BOTTOM*). The photos were taken at the same time and from the same spot. To understand this better, refer to the photo on p. 171.*

CLOCKWISE FROM ABOVE: *Jared with Jill, BHU, 1990–91; parsnip flowers among the apple trees; working in the BHU orchard, 1990–91.* (Photos courtesy Jared White)

The Project Gro sign listing all the donors disappears under fennel, comfrey and grasses, November 1999.

Autumnal splendour of Bob's home garden, with Stevenson screen and rain gauge in the lawn.

2009 BHU calendar showing impressive revival. Images from 2008. (Courtesy Heather Hughes)

CLOCKWISE FROM ABOVE: *Bob with delphiniums; Madge enjoying coffee and cake in Bob's garden, 2000; last coffee morning at Ashgrove Terrace, 2023: Anne Simpson, Geoff Barnett and Bob; home garden, Ashgrove Terrace.*

Bob, 2023. (Photo: Cullen Pope)

France introduced organic farming into government policy in 1980.[49] In the US, the United States Department of Agriculture (USDA) released the Bergland Report in 1980, which concluded that 'organic farming would receive an impetus from increasing concerns over energy shortages, declining soil productivity, soil erosion, chemical residues in foods and environmental contamination'.[50] The last of the report's 19 recommendations reads: 'It is of utmost importance that USDA develop research and education programs and policies to assist farmers who desire to practise organic methods.'[51] Bob was gradually becoming aware that the world was beginning to embrace organics.

•

Things were also dynamic at a domestic level, although in a more prosaic sense. Bob's partner Chris wanted a dog and guinea pigs, so they moved from Carlton Mill Road into a rental property in Cashmere Road and acquired Jill, a German shepherd.[52] The house was destroyed by fire while Bob was in Britain in 1979, and the Morris side helped Chris to move their belongings to a temporary rental in Colombo Street. Chris then began to look for a more permanent solution and eventually found just the place.

The couple visited 73 Ashgrove Terrace together in spring 1980 and found it 'a floriferous wonderland'.[53] The area between Ashgrove Terrace and Rose Street had extensive flower and market gardens at the time, and the residential properties were mostly half-acre (2000m²) blocks.[54] Horses grazed on either side of number 73, which had a stable housing chickens. It was indeed the perfect place, and Bob bought it. Bill was delighted: 'You have now become a gentleman of estate with your own homestead. I think you have a very nice little place … somewhere you can dig your heels in and develop. It will be nice to see how the garden has improved after you have had it for a couple of years.'[55] At the time Bob was 41, and 41 years later he still lived in the same house. The gardens of Ashgrove Terrace are a significant part of Bob's story and have numerous associations. One of these, the fig tree immediately outside the kitchen, came from a cutting of the tree Bob had planted at Carlton Mill Road, which in turn came from his friend Leon Langley.

Chris was an active participant in the Morris side and sometimes assisted Bob with harvesting and other tasks at Lincoln. But despite their many overlapping interests, he soon left Bob for another man and, after several years, left the country altogether. Bob became settled, more focused and happier: he had his dream house, a dog, and a strong spiritual life as expressed through Morris dancing.[56]

Ashgrove Terrace viewed from the back of the property in 1980 (top), and the same view several years later showing the garden development (main picture)

Large-scale compost making at the BHU.

The one area of his life that lacked resolution was his work. Was organics worth throwing his professional weight behind? It was interesting, but not without challenges. In these 'transition years' the produce from the organics plots at Lincoln was poor: crops were 'hit by pests, diseases and nutrient deficiencies. I kept thinking – oh, God, is it worthwhile persevering?'[57] The push-back from some colleagues had turned to scathing criticism. As Bob said, 'It was a time of people thinking you were a bit cranky if you wanted to do organics.'[58]

Bob had shifted from merely allowing keen students to 'do their thing'. He now actively built organics into the course work. He trod carefully: he did not overtly promote organics to the students, but required them to compare organic and conventional systems within their plots – double digging with minimal tillage, for example; transplanting with direct sowing; or Nitrophoska Blue with bone dust or compost. They used the organic plots that Tim Maples had set up and made compost heaps, but this in itself was hardly radical: 'I had to be pretty careful that I didn't cross the line, so that the students could not say I was indoctrinating them with alien concepts that did not meet the academic perspective of the university.'[59]

From October 1980 he began to sense his calling. He wrote to his friend Tony West, saying his 'own feelings on what is happening are becoming increasingly clear and my whole research effort and teaching will become increasingly environmentally orientated. In fact, I shall even be considering getting out of Lincoln when the right opportunity arises.' The new house had created opportunities: he wrote that his half-acre garden was coming along and he had already 'had a local environmental group along for discussions and practise, and hope to develop the area as a teaching aid and example of how to go from chaos to order with minimal tillage and capital investment'.[60] By September 1981 he had established a Biological Husbandry Group 'with the aim of promoting more actively our beliefs', and in early 1982 he was able to say that he had been 'working slowly … to increase the awareness of students regarding environmental aspects of life and … running an organic garden now into its fifth year'.[61] He pointed out to colleagues in the UK that he had 'some forty final-year degrees working on a critical appraisal of the research done to date on alternative agricultural techniques'.[62] Far from being scandalised, the students found this work stimulating.

So did Federated Farmers, who approached Bob as a potential keynote speaker for its conference in early 1981. It requested Bob's synopsis and a copy of the influential USDA Report on Organic Agriculture, the Bergland Report, 'without delay'.[63] Bob saw this as an exciting challenge 'and an excellent incentive to get one's thoughts into line'.[64] He proposed to call his talk '"Think Big"? Naturally' – 'in deference to our great leader': Prime Minister Robert Muldoon.[65] He pondered whether to use the terms 'organic farming', 'biological farming' or even 'environmental farming' – nomenclature that would be debated for years. He would explore recent global developments in organics, share some of his own research on tomatoes and gherkins, and describe New Zealand's opportunity to mark out a niche, capitalising on 'our one asset, purity in a polluted world'.[66]

In the end Federated Farmers decided not to proceed with Bob's talk, but chief executive Rob McLagan commented, 'No doubt the subject of biological husbandry will become of increasing interest and importance and we may therefore be in contact with you again to seek your assistance in introducing the subject for discussion at a future federation meeting.'[67] That opportunity came the following year.

The delay probably worked in Bob's favour. He was aware that he needed 'a greater understanding of the scientific credibility of organics', so he self-funded an international study tour in April and May 1981.[68] He visited British Soil Association's Walnut Farm, and Elm Farm, its soon-to-be replacement. He went to Santa Cruz (already familiar to him from his previous Californian visits), and the Organic Garden and Farm Research Center (funded by Rodale) in Pennsylvania.[69] He visited Witzenhausen in Germany where he met Professor Hardy Vogtmann, whose influence on the international organic farming movement was already profound. One of Hardy's master's students then took Bob to Obervil in Switzerland, the biodynamic research centre where Hardy had begun his career.

The trip 'confirmed the direction that I was going at Lincoln ... I came back and felt that it was the way to go, and I was justified in pressing on with what we were doing.' As he wrote at the time, it heightened his resolve 'to try and influence the direction of research and development in agriculture' in New Zealand.[70]

Bob believed New Zealand would benefit from joining IFOAM; he himself joined as an individual on 9 September 1981. His letter to IFOAM stated that for five years he had been developing 'a small area for organic production ... intended to stimulate interest of the students in organic food production and hopefully to develop into a meaningful research facility for organic husbandry'.[71]

By 1982 Bob's work on organics at Lincoln was attracting a number of interested people. One of these was Tim Jenkins; like Geoff Barnett before him, he had come to Lincoln specifically to meet Bob and would later play a critical role in the Lincoln unit. Tim was already interested in organics and had attended a hui on the topic in 1981. His father was a neighbour of one of Bob's Lincoln colleagues, Trevor Jackson, who arranged for Tim to meet the man himself in 1982. Tim was only 14 at the time: 'I remember it well, seeing the permanent cultivation-free beds, the large compost heaps and visiting the row of youngish pear trees that had just had compost applied to them ... I was soaking up what he was explaining for the vision of organics.' In 1986 Tim returned to Lincoln 'precisely so that I could study with Bob and have an aspect of organic and sustainable farming'.[72]

The term 'biological husbandry' was increasingly used as an alternative to 'organic farming'; as early as 1979 *Soil & Health* reported on the commercial viability of 'organic or biological farming', and in 1980–81 the publication featured a section on 'biological farming' with articles such as 'How organic or biological farming compares with chemical farming'.[73] Tauranga Community College was running a course in biological husbandry in 1982, and the Soil Association stated that it preferred 'the use of the term Biological Farming [to organic] as the term is being adopted throughout the world'.[74] It was helping the organisation gain the kind of credibility in mainstream farming circles that it had not had since the 1940s. At the same time, however, the members of the Soil Association had voted for the words 'organic growing' to be emblazoned on the cover of its magazine.[75] A growing divide was forming in the membership between the ('organic') gardeners and the ('biological') farmers.

The Soil Association was charting a new course. Mollie Chalklen became the association's national president in 1981.[76] Jack Meechin retired as magazine editor, a position he'd held since 1954, and *Soil & Health* underwent a massive facelift. Mollie was keen for the organisation to modernise and engage more directly with new opportunities as they appeared. One of these was the development of an organic farming sector, and Mollie was very focused on this. She knew of Bob's work and had spoken about his research on National Radio in May 1981.[77]

Mollie believed the Soil Association had been poorly managed by the Christchurch-based committee. It was, in fact, 'broke'.[78] Her strategy for renewal included shifting the head office from Christchurch to Auckland and ensuring that younger people with sound financial skills and an entrepreneurial mindset were at the helm. At the 1982 AGM in Dunedin she met Chris May, a young West Auckland (Waimauku) farmer who was attending in his capacity as chair of the Auckland branch.[79] Chris had exactly the qualities she thought the association needed. His proposal at the AGM – that an umbrella organisation for a New Zealand organic certification system be established – won Mollie's approval, and the association awarded the Auckland branch $500 to begin the project.[80] Chris and his wife Jenny, with Dave Woods – essentially a subgroup of the Auckland branch – began formulating the concept of what would later become a Biological Producers Council.

A month after the Dunedin AGM, on 11–12 June 1982, Chris, Jenny and Mollie attended the landmark Seminar on Biological Farm Management Techniques convened by Federated Farmers at Flock House in the Manawatū.[81] The seminar was the outcome of the discussion that had begun between Bob and Peter Waugh of Federated Farmers the year before. It aimed 'to bring together those people already involved in biological husbandry, and those interested in establishing such techniques for their own farming operations and for research purposes, to assess what areas [are] worthy of research and by whom'.[82]

The event opened with an address from Simon Upton MP and speeches from organic farmers John Scott, Michael Templer, Marinus La Rooj and Ian Stephenson.[83] Bob's talk was listed in the programme as 'The scientific credibility of biological husbandry' (the write-up in Federated Farmer's journal *Straight Furrow* called it 'Can we think big – naturally?').[84]

The Flock House event marks a turning point in New Zealand's organic farming history, and Peter Waugh's role in making it happen appears to have been pivotal. Peter had recently returned from studying Naturecure (a precursor of naturopathy) in India and was involved with a new organic co-op in Wellington. He fell into a role with Federated Farmers as legal advisor and secretary of its Rural Land Use Committee.[85] He recalled his interest was in 'human health, and to see sustainable farming methods being used – and part of that is organic farming methods'. Others on the committee 'saw the reason for exploring where things were up to in the whole field of organics in New Zealand, and where it could potentially go'. At the time, very few farmers of scale were 'doing it organically ... They were regarded as really fringe elements, because there was no market for organic meats or anything at that stage ... There were idealists growing organic vegetables, but still there wasn't really a hell of a lot of market for those.' Peter had visited some of the farmers who were using organic principles and had come across Bob in 1980.[86]

Federated Farmers was 'wanting to broaden their perspective,' Peter recalled. 'There were some very enlightened farmer politicians involved, such as [Sir] Peter Elworthy ... And they were open to looking at alternatives.' Since the late 1970s, organic farming had been gaining a 'heightened stature among researchers, educators and agricultural policy makers', and Federated Farmers agreed that backing a conference of this sort was within its purview.[87] 'The idea behind it was to bring into the open what was there in the way of research, and researchers, and people practising – to bring those people together with practising farmers.' The response was overwhelming. The Flock House event included speakers from Massey University, manufacturers, fertiliser company representatives and people from the farming community, and was an opportunity for researchers, who normally could not be heard, to present what they had to offer. Peter recalled that the organic movement in New Zealand was 'very fragmented'; the seminar on biological farm management techniques brought people together and received a write-up in a special 'Organic Farming' supplement of *Straight Furrow*.[88]

In his presentation to the seminar, Bob stated that he considered the term 'horticulture' to be 'an anachronistic concept ... intensive cropping of any description belongs within the framework of a sound fundamental agricultural system'. He outlined some of the major findings from overseas research stations such as Elm Farm Research Centre in Britain, and from William Lockeretz at Rodale Institute in the US and Miguel Altieri, also in the US. He explained that the organic movement was global and that 'the co-ordination of this worldwide development has been by a voluntary organisation known as IFOAM (International Federation of Organic Agricultural Movements)'. He concluded on a note that evoked his deeper commitment:

> Biological husbandry is a total environmental concept that embraces all things, the emotional satisfaction of a nurtured landscape in total balance with the need to produce both for self-sufficiency and to maintain quality exports where the name of quality is purity.[89]

Mollie was anxious that Bob, Chris and Jenny should meet and introduced them to one another at Flock House.[90] The move was wise and prescient: Bob was obviously interested in what Chris and Jenny May were doing.

Bob attended a meeting in Auckland on 9 August 1982, called to form an Organic Growers Council of New Zealand.[91] The invitation came from the Doubleday Research Association, the Soil & Health Association, the Biodynamic Farming and Gardening Association, the Biological Husbandry Group (at Lincoln), the Permaculture Association of New Zealand and others. The meeting was the follow-up to work initiated by Chris and Jenny. They invited Bob to speak and flew him there from Christchurch. 'Bob was very important in giving people around him credibility because he was from Lincoln,' Chris said.[92] Involvement in this national discussion was good for Bob, too, as Jenny noted: 'I think he had the same vision for the world as we did … it was a mechanism for sharing that vision.'[93] They liked Bob and were 'very defensive of him'.[94]

The activity since May had been dizzying: the Soil & Health AGM in May that endorsed the focus on developing a commercial organics sector; the Flock House seminar in June; and an Organics Growers Council initiated in July. This was the birth of a commercially savvy organics sector in New Zealand. For Bob, it may have been too much all at once. At some stage he was struck with a debilitating 'flu and believed that he never fully regained his energy.[95]

In August Federated Farmers, again through Peter Waugh, asked Bob to provide his ideas for research priorities.[96] Bob's reply to Peter is important: research, he said, should be carried out on properties already using biological husbandry principles. He noted that some of the scientists at the Flock House event 'could not accept the presented data and generally commented on the lack of acceptable data to back up claims'. He believed that with a small injection of funds, rapid results could be gained regarding the economic viability of biological husbandries of the sort seen by William Lockeretz et al. in the US Midwest, where organic farms had produced crop yields and net farm income similar to those of conventional farms.[97] Bob believed such work would highlight further research needs and help to determine 'whether there is a need for the establishment of … a Biological Husbandry Unit or units attached to a specific research establishment'. He pointed out that the 'Biological Husbandry Unit' at the Horticulture Research Area at Lincoln College had been in operation for five years.[98] This appears to be the first time he referred to the organic section at Lincoln as the 'Biological Husbandry Unit' – until then it had been the 'organic gardens' or 'an organic area'.

Bob's sense of purpose was clear: he saw that he could work with Federated Farmers to encourage them to do the preliminary work to underpin the development of a proper organics research station along the lines of those he had visited the previous year. That trip, the interest in research from Federated Farmers and the development of a clearly delineated organics sector had helped to crystallise the

Bob in the tunnel house, Horticultural Research Area, c.1982.

idea in his mind. As well, the unit at Lincoln was set to expand: he was about to take over additional land, including an orchard developed by David Jackson and some plastic growing tunnels.[99]

Mollie was delighted by all this development, as she exclaimed in 1983:

> Exciting things are happening in the Soil Association. The seeds sown so many years ago by members convinced of the importance of organic farming are at last growing sturdily and beginning to bear fruit. The Seminar in Biological Husbandry last June spurred interest in alternative methods of farming. Since then, visits have been made to organic farms by Federated Farmers, DSIR [Department of Scientific and Industrial Research] and MAF [Ministry of Agriculture and Fisheries] who have been impressed by the results obtained, and have asked for more information.[100]

Mollie was referencing the outcome of Bob's suggestion to Federated Farmers: a significant research project led by MAF and – with Bob's support – a detailed literature review commissioned by Federated Farmers through Peter Waugh. Written and edited by Richard Hudson and John Calvert, *Ecological Agriculture: Review and annotated bibliography* (1983) was important in outlining a large collection of scientific research into biological husbandry, which the authors defined as 'an alternative to conventional industrial agriculture. It incorporates a vast range of techniques and practices which aim at avoiding pollution and ecological problems, and to produce healthy food of high nutritional value, whilst sustaining the means of production, the soil, air and water to a high-quality standard.' They believed that biological husbandry was 'now a long way beyond trying to prove its worthiness beside conventional agriculture' – although it was still very much at that point in New Zealand.[101]

In her autumn presidential column in *Organic Growing* (as *Soil & Health* was briefly renamed), Mollie Chalklen referred to a second seminar on biological husbandry to be held at Lincoln College in May 1983, saying, 'It is good to continue working with Federated Farmers and for our growers to be with the scientific fraternity on a research committee.' She reminded readers that the Soil Association offered a $1000 postgraduate fellowship for research into biological husbandry at Lincoln and noted that at Lincoln two new papers in biological husbandry were anticipated to count towards a professional qualification. This was a lot of news to lead on, and although she didn't mention Bob's name, his influence was apparent.[102]

The theme for the 1983 Soil Association conference, held in Auckland in May, was 'Looking at Commercial Organics'.[103] Bob was a keynote speaker. At the conference Chris May was elected president. The conference supported the plans Chris had been working on for a Biological Producers Council that would, as he wrote, 'offer both consumers and growers a confident base from which to buy and market produce':

> For 40 years the home gardener has benefitted from the leadership and information provided by the association. This important role will continue and will be encouraged. At the same time, it is obvious that we must broaden our base to give more support to the people who make their living from 'Growing Organically'.
>
> … For so long we have talked about offering the Ministry of Agriculture and Fisheries advice or writing to the Minister with suggestions. It is far more important to lead by example. Much of the credibility of biological husbandry lies in its economic viability. That is why farmers like John Scott, Ian Stephenson and John Pearce are now visited by interested groups.[104]

In her speech as outgoing president, Mollie reflected on this tension. She was delighted by the success of the 1982 seminar on biological husbandry and the interest since shown, in particular by Federated Farmers. But she observed, cannily, 'we must not forget the basis on which we rely for support. The great majority of our members are home gardeners: without them we would not be here today.'[105]

Chris May's enthusiasm for pursuing the commercial opportunities of organic farming meshed perfectly in timing and approach with Bob's teaching and research at Lincoln. Working together, they were able to support one another's missions. The Soil Association brought a growing community mandate and funds to Bob's programme; Bob's work brought vital scientific rigour, credibility and institutional networks to the Soil Association.

At the Auckland conference members voted in support of the proposed New Zealand Biological Producers Council. Former long-standing mayor of Auckland and one-time president (and now patron) of the Soil Association Sir Dove-Meyer Robinson spoke, and guest addresses were also made by the former minister of agriculture, Colin Moyle, and Minister of Agriculture and Deputy Prime Minister, Rt Hon Duncan MacIntyre. MacIntyre spoke of the interest generated from the Flock House event the previous year and his hope that organic farming could offer a solution to current challenges by 'cutting costs and boosting outputs'; Moyle stressed the importance of all New Zealand producers ensuring that 'their produce carries the "connotation of purity"'.[106]

Bob's own address was a solid case for the commercial viability of organic systems, outlined in his first written contribution to *Soil & Health* later that year, titled 'A guide to organic growing profitability'. He spoke about crop yields, disease and pest incidence and inputs (but not labour costs) observed over six years of cropping at the 'Biological Husbandry Area' at Lincoln College, a subset of his Horticultural Research Area. Couching his words carefully, he concluded that the 'values given do indicate that production in general on the plots is at a very acceptable level as is the quality in most cases'.

> The Biological Husbandry or Organic movement has always been based on a strong moral and ecological footing. Increasingly, it has become based on a strong economic footing and this is why the farming community is so interested in what is going on within the movement today.
>
> Let us capitalise on the interests of the community and go forward with strength and energy to create a new system of agriculture based on a return to Biological Husbandry.[107]

The Auckland conference was followed immediately by the second seminar on biological husbandry at Lincoln College, which was reported in *Soil & Health* under the heading 'Lincoln College man's view'.[108] The author stressed that they were quoting from a Ministry of Agriculture report in which Bob was reported as saying that with 'increasing worries about agricultural pollution and the oil crises, there has been a lot of interest in a system of farming that has low energy inputs and is ecologically sound'. He listed the aims of biological husbandry as: maintaining long-term soil fertility, avoiding all forms of pollution resulting from agricultural techniques, producing foodstuffs of high nutritional quality, and reducing energy inputs to a minimum. 'The system aims to provide a balanced environment in which the maintenance of soil fertility and control of pests is achieved by enhancing natural processes, not by working against them.'[109] In the same report, farmer John Scott explained his experience of nearly two decades of organic growing.

Bob maintained later in life that he didn't have 'a blinding insight into the strength of natural systems. I had a blinding insight into the failure of industrial agriculture ... and the energy crisis.'[110] However, he was clearly impressed with the functional biological systems that he had observed and perhaps none more so than the extraordinary example of John Scott's property.

Around 1983 Bob arranged a field trip for his students to visit John's 120-hectare Hawarden farm. There was a high level of interest. The farm had transitioned from wheat cropping and burning in the 1920s and 1930s, which had led to problems with erosion. In 1937 John's father took over and introduced a policy of no cropping – fodder and hoggets only – and probably used superphosphate. He was bought out in 1949 and the farm was used for dairy, sheep and cropping. John bought the property in 1965 and introduced an organic regime. His foundation was pastures with a mixture of species to draw on the total soil fertility, and his philosophy was 'biological balance, not control'.[111] John's property was a living example of what an organic farm could look like after nearly 20 years.

A legitimate market for organic food was beginning to open up. Bob was already on record talking about the growing market for biologically produced food, particularly in the United States.[112] In Christchurch, Piko Wholefoods had opened its doors in 1979, and there was the new organic co-op in Wellington that Peter Waugh was involved with.[113] Bob believed the 'days of saying that eating organically grown food was okay if you didn't mind holes in your apples and grubs in the lettuces' were well gone.[114]

A third seminar on biological husbandry was held at Lincoln in July 1983.[115]

The Soil Association's Canterbury branch put on a massive exhibition of organic produce at the Canterbury A&P Show in November 1983. The theme for the central exhibit that year was 'Alternatives in Agriculture'. Lincoln traditionally had a space at the show but had no alternatives to promote, so Bob offered the space to the Soil Association and drove the project. The committee's theme was 'Biological

Organic display at the 1983 Canterbury A&P Show.

Husbandry: The Organic Alternative', and the team made sure that scientific facts accompanied the many images on display.

The Soil Association exhibition included space for the 'Biological Husbandry Demonstration Unit at Lincoln College' and a section from Trevor Jackson of the Ministry of Agriculture Research Division – Entomology. In the biological husbandry section, where signage read 'Don't sell our future down the river', a binary was drawn between 'poor' (conventional) farming and 'good' 'biological husbandry'. The science and international section of the exhibit was entitled 'Think Big – Naturally' – a return to Bob's earlier theme. Bob hoped this section would show the public 'that the Biological Husbandry Movement is not just quaint folklore, but a growing, dynamic world movement based on seeking a true understanding of the workings of the biological system'.[116]

The display brought together produce from all over the country and showcased top-quality vegetables grown at Lincoln's 'Biological Husbandry Demonstration Unit' – the name Bob had started using for his research area. The impressive collection of organic produce featured on the cover of *Soil & Health*.[117]

Bob had announced his own 'conversion'.

Bill, Madge and Bob sight-seeing in the Waitakeres, 1983.

5: Developing organics, 1984–87

BOB'S ACTIVE SOCIAL LIFE centred around Friday evenings at The Landing, Leon Langley's restaurant in Cashel Street. The Landing was not Leon's first venture: he had opened the Attic Coffee House in 1957, described as one of the first modern cafés in Christchurch.[1] In 1979 Leon opened the Green House on Colombo Street above Ballantynes, a gay-friendly part of the city (a men's sauna had opened up the same set of stairs), and it was here that Bob came to know Murray Scott, whom Leon employed as a musician. They started talking in Murray's breaks and 'one thing led to another'.[2]

Soon after this meeting, Bob's parents arrived on their first visit to New Zealand. They would have come earlier, but Madge had refused to put their ageing dog, Shandy, in kennels. In 1979 Madge had been worried about 'the Unions', electricity cuts and rising prices: 'I think NZ must be better even if you say it is going down the drain. If GB goes any worse, I think we will sell up and come out and buy a bungalow.' By 1980 she was even more down-hearted: 'There is nothing I would like to do better [than come to New Zealand] but I can't see it happening in the near future. Shandy still flourishes and I can't see Father leaving her.'[3] Shandy died in April 1982, and at last the couple were free to travel. They stayed for three months over the summer of 1982–83, just as Bob and Murray's romance was blossoming.[4] If Bill and Madge were aware of the relationship, they never discussed it with Bob.

•

In 1983 New Zealand's organic movement had gathered some momentum, but 1984 began with a blow. MAF released its 'Report on Biological Farming', which looked at three organic farms and two biodynamic farms, and compared one organic farm and two biodynamic farms with three conventional farms. Despite noting that animal health, economic performance and the quality of produce were all better on the organic farms, the report's authors were sceptical about the value of organic practices. The report was reviewed unfavourably in *Soil & Health*. Valerie Thompson wrote of how disappointing it was to read that the authors believed conventional farmers already used organic practices, including the use of legumes to fix nitrogen, cycling of nutrients through grazing animals, and liming soils to encourage earthworms. 'In fact, they see the main difference between organic and conventional farmers as being the use of slow-release fertilisers instead of quick release ones, and the lack of use of animal remedies.'[5] Perry Spiller later wrote that the report 'was roundly and soundly condemned by all organic farmers, including those whose farms were surveyed.'[6]

Bob was among those who were not impressed. Commenting on a meeting of the Steering Group for Biological Husbandries held in mid-December 1983, he wrote in *Soil & Health*: 'It is now over 18 months since that memorable seminar at Flock House, June 1982, set the scene for Biological Husbandries in New Zealand under the driving force of Peter Waugh of Federated Farmers. In that time, we have come quite a long way, or have we?'[7] At the December meeting the Minister of Agriculture had 'made a token visit for lunch with the committee': 'Congratulations and back slapping were the order of the day.' The DSIR entomology representative 'was at pains to make clear that their programme of biological and integrated control had nothing to do with the current wave of interest in biological husbandry'.

Of the report itself, he was scathing:

> The results of the Survey of Biological Farms carried out by the Ministry of Agriculture showed: 'No striking differences between the chosen farms and comparable conventional farms.' Not a sound about this at this meeting; a self-satisfied silence. But if that is not a 'striking' result, what is?

> … How long must we sit around while much of the scientific community plays around trying to convince themselves that they, like the entomologists of the 1960s, are following the wrong direction in the management of our heritage and welfare?

> … One wonders if the people on the committee have ever read the literature that has constantly been fed to them over the last 18 months, because the understanding of the meaning of Biological Husbandry still appears confused.

> We are talking about a system that is self-sufficient and will stand on its own. We are not talking about using certain aspects of Biological Husbandry to prop up an agriculture that is at present propped up by the taxpayer.[8]

Bob was concerned that this approach from 'establishment' would quell 'the vim, verve and vitality of the existing movement'. He made an argument for New Zealand-based research into 'making the Biological Husbandry "system" more efficient in every respect, for example at the Biological Husbandry Demonstration Unit at Lincoln': 'There is no doubt we have to Think Big – Naturally, and independently, if we are to make real progress.'[9]

Chris May agreed with Bob on this last point:

> Independent Research into Biological Husbandry is vital to the organic movement as a whole. Although there are many individuals growing organically, we need to work closely with an institution that provides a sound basis through research from which organic growing can be encouraged and helped scientifically.
>
> The Lincoln College Biological Husbandry programme under the direction of Mr Bob Crowder, Senior Lecturer in Horticulture, provides us with a point for focusing our energies in the research area.[10]

In May 1984 the Soil Association held its annual conference in Christchurch, once again opened by Sir Dove-Meyer Robinson. The conference included visits to Bob's experimental five-hectare organic plot, which he wanted to expand to 10 hectares. The plot, described in *Soil & Health*, was 'designed to demonstrate a functioning organic system of growing vegetable crops':

> In his experimental plots at Lincoln, Mr Crowder has formed a broad-based mix of vegetation including fruit, berries, culinary herbs, flowering shrubs, trees and garden flowers as well as vegetables. The fruit trees in the unsprayed experimental garden are underplanted with a mixture of herbs which will not only provide a home for predators but also suppress the grass which is undesirable in an orchard. The ground-floor herbage in an orchard should be winter-green and dormant in summer, when the dead tops provide a mulch. Some of these plants are often regarded as weeds. Others are onion flowers, grape hyacinths, comfrey, foxgloves and daffodils, all attractive companions for fruit trees.[11]

At the conference, Dove-Meyer Robinson formally announced the Soil Association's launch of Project GRO (Giving to Research in Organics) for mid-1984. Project GRO – a crowd-funding campaign that allowed members to make direct financial contributions to key projects – would provide funding for the Biological Husbandry Demonstration Unit at Lincoln to expand its research programme and establish an organic advisory programme, and would support a full-time assistant to help the research scheme.[12] Chris May regarded the project as 'a major priority for the Soil Association – our credibility and progress lie in research into Biological Husbandry'.[13] This point is critical: the government's failure to provide an adequately researched profile of organic farming had pushed the organic movement to take matters into its own hands; in doing so, the movement positioned Bob as a man of the people rather than a man of the establishment.

Bob was relieved to have a meaningful outlet at last for the frustrations he had felt with the horticultural world in which he worked. 'Like many lecturers at Lincoln College, I think I also felt I did lecture basic fundamentals of production … But on looking at the outcome of my efforts, I realised that I, like many others, was doing what I could rather than what I should.' He emphasised that the development of the demonstration plot in 1977 had come about as a result of pressure from students, 'not with a fanfare or request for establishment funds, but a slow subversive infiltration into 0.2 hectares located at the far extremity of the research area, out of sight, out of mind and established with the help of students and interested persons to demonstrate an alternative system of production.'[14]

The demonstration plot was now transitioning into a research area: Bob was striking out on his own within Lincoln College. The Biological Husbandry Demonstration Unit would, he said:

> continue to develop the concept of an alternative philosophy or system based on a self-sustaining strategy for survival. As the area passes from a demonstration unit to a full research unit, investigations will be conducted into aspects of production from within the confines of the biological husbandry system, rather than using biological husbandry techniques within the framework of the conventional system.[15]

Such comments were not calculated to curry favour within the institution, but were crowd-pleasers for the organics movement.

Bob's manager, Mac Morrison, continued to support Bob's work. Sabbatical and conference leave was leveraged to allow Bob to travel to Wellington to help develop BioGro, the organic certification programme run by the New Zealand Biological Producers Council, and attend IFOAM conferences. Another source of funding for his work came from the British branch of the C. Alma Baker Trust, which was established to further the science of agriculture. This funding was orchestrated by Perry Spiller, Chris May's successor as president of the Soil Association, who was disappointed that the New Zealand chapter of the trust had not been forthcoming.[16] Bob tended to travel cheaply. 'All my work has been basically courtesy of people putting me up in beds and floors all over the world rather than having to stop in a hotel, which I would never have been able to afford.'[17]

Despite tacit support from his manager, Bob experienced a growing feeling of ill will from some of his peers. Bob Douglas observed that the work of bringing organics to students was 'a struggle'. Funding was difficult to source, and Bob 'came up against a lot of resistance' from the academy and some sectors of primary industry.[18] In Bob Douglas's eyes at least, this antipathy built upon an existing prejudice that pitted rural agriculture students against urban horticulture students. It was, he said, 'the resistant attitude that the typical Lincoln student had towards the horticultural students … that sort of solid, right [-wing], entrenched attitude'.[19]

Parsnip flowers in the BHU apple orchard, a part of the 'floriferous, umbelliferous understorey' for biodiversity enhancement. In this system, parsnip flowers always followed a riot of white cow parsley flowers.

But Bob recognised the extent to which he could push things to leverage the resources at Lincoln to nurture young people in their aspirations to change the world, and to further the immediate aims of the wider organic movement. He was highly successful at this.

It was obvious to many horticulture students that Bob was treated poorly by his colleagues. Brendan Hoare, who came to Lincoln as a 19-year-old in 1984 to complete a Diploma in Horticulture, was one such student. Brendan had been doing a cadetship in horticulture in the Gisborne area and chose to focus on organics, but was told 'that organics was nothing to do with horticulture'. When he failed the cadetship, one of his tutors told him about Bob and recommended he go to Lincoln. Brendan wrote explaining the situation and Bob encouraged him to come south. He rode to Christchurch on his motorbike with his surfboard on the side. Bob, he said, was 'hammered' by his fellow academics. '[He was] laughed at … When I got to Lincoln there were people who used to ridicule [Bob].' It seemed to Brendan that there was more to this derisiveness than Bob's method of practising horticulture: 'It was clear Bob was gay … That snide remark thing … it was always just there … in my classes I would always stick up for organics and ask questions that used to piss the lecturers off, partly to support Bob.' Bob's combination of being gay and into organics and Morris dancing may have been too much for the Lincoln establishment, but to Brendan he was 'full of life … and so courageous'. There was also Bob's generosity, which came in many forms, including 'his famous coffee and cake sessions' in his 'fantastic garden':

> Bob was such a good teacher because he … enabled me to polish what I wanted to do … The cadetship I did made a fool of me, and I found this Morris dancer … I thought, man, if this guy can …. That is what inspiration does … It enabled me to be out there … Bob always had this wonderful sensitivity to things, flowers and things that are tiny, and I'd always had that too, so he was a kindred spirit in that way – and that's crucial when you're young and in your formative years.[20]

Brendan became part of the biological husbandry group who volunteered at the unit and occasionally attended Soil & Health meetings. 'Even though we were a tiny group, I knew we were right. [Bob] was really inspiring in that regard, how he stood up for what was right … I wanted to change horticulture and the way that we treated Papatūānuku … Bob was the first one who gave me courage.'[21] Brendan attended meetings of the Canterbury Organic Producers (COP) group, which included local organic farmers such as Tony Mallard, Ian Henderson, Rose Donaghy, Tim Chamberlain and Ernst Frei. When COP began is uncertain – presumably after BioGro started in 1984 – but by 1986 it had 50 members and a year later almost 90. (By 1990, though, numbers had declined to 45.)[22]

In 1986 another student, Joanne Blakely, co-authored the report 'Organic Horticulture' for the Horticultural Market Research Unit, correcting some of the ideas set out in the ministry's report three years earlier. In *Soil & Health* she wrote,

> My personal involvement with biological husbandry stems from my Lincoln College days where all horticultural science degree students were introduced to organic growing principles by Mr Crowder. I finished my degree in 1982 with the understanding that our horticultural industry is dependent on chemicals for its survival as an export earner, and I concentrated on learning more about integrated pest management programmes where growers resort to chemical control [only] when the pest or disease reaches an 'economic threshold level'.[23]

Joanne became a junior lecturer in horticulture management at Massey University, where she set up the first course in organic farming systems in the Agricultural and Horticultural Systems Management Department: 'They allowed me to create [a] paper … which looked at all different components from soils right through to examples of different systems that were running.' Joanne left Massey in 1989, at which point some of the momentum was possibly lost.[24]

Another student who appeared in 1984 was 24-year-old Jon Manhire, who was doing a Postgraduate Diploma in Horticultural Science. Like Brendan, Jon became involved in the student-led organics group at the unit. At the time Bob was fired up about the development of BioGro, and he shared his excitement with these students after each successive trip to Wellington.[25] By 1987 Jon was working as an agronomist in Timaru, where he started up Intrepid Seeds with David Musgrave in 1989.[26] Shortly after this he commenced working for MAF, where he led much of the government's organics programme.[27]

The students also helped Bob to identify his own beliefs. In 1984 Bob wrote:

> I started to think, in company with my students, that we were perhaps not teaching science – horticultural in particular – in the correct manner … With their help we collected a great amount of literature … I felt that the world was in a really difficult situation and that the insidious pollution of the environment is a very much greater threat to the survival of the world than the nuclear holocaust. And I felt that perhaps New Zealand was a country that could set an example to the rest of the world on how an environment should function.[28]

> … To change to biological husbandry, we have to break through a philosophical barrier. We change ourselves. From wanting to dominate nature and telling it where we want it to go, we move to where we want to integrate with nature, work with it and let it do the work. In biological husbandry we find strength through diversity.[29]

Geoff Barnett harvesting summer vegetables, 1982.

Project GRO enabled the employment of a full-time technician for the unit in 1984. Geoff Barnett first met Bob in 1977, and his appointment to the role was announced in *Soil & Health*. Since completing his Diploma in Horticulture at Lincoln, Geoff had 'been gathering practical experience on farms and other properties in the Canterbury district'. He came to the Biological Husbandry Unit as a labour scheme worker, assisting with general maintenance during the 1982–83 season, and had 'proved himself very suitable to be offered the first, permanent position'.[30]

Geoff was involved in the intensive beds initially set up by Tim Maples. He made a lot of compost, helped students to manage their plots and took care of them while students were away on holidays, sometimes with the help of summer interns. Geoff did not need much encouragement regarding organics – 'it just made sense to me' – and he continued as technician until 1998, broken only by a two-year absence when he went overseas.[31]

As Bob became more marginalised at Lincoln, his connections with the wider organic movement became increasingly important. He attended his first IFOAM conference in 1984 at Schloss Ludwigstein and nearby Witzenhausen in Germany, where he arranged a display of goods produced at the Biological Husbandry Demonstration Unit and presented a poster session describing its development.[32] The conference was sponsored by IFOAM and the University of Kassel, which was home to the Division of Alternative Agricultural Methods, established by Professor Hardy Vogtmann in 1981 within the Department of Agriculture.[33] Bob found the experience 'epic'. He was particularly inspired by Hardy's accomplishment: in three years he had turned an extensive farm into a research centre that was 'totally involved with investigating aspects of the biological husbandry movement using the energy input of undergraduate, masterate and PhD students'.[34]

This was exactly the vision Bob had for his own area at Lincoln. 'It was just as important for me to go to [the conference] as it was to go to an international convention on horticulture'.[35] The conference, he felt, 'certainly showed that the biological husbandry movement is not sustained by just a bunch of quaint eccentrics' – perhaps a reference to the reactions of his Lincoln peers.[36] It was, rather, 'the Organic Movement, new style, a reflection of the growing concern of many people from all around the world for the reckless exploitation of the world's resources, not only in the developed world, but also in the Third World'.[37]

At the conference Bob heard from many big names in the organic movement, possibly for the first time. These included Hardy himself; Miguel Altieri, an agroecologist from the University of California; Lawrence Woodward of Organic Research Centre Elm Farm in the UK; and Garth Youngberg, who had played a critical role in elevating the status of US organic farming while in the USDA.[38] Youngberg spoke of the improving policy environment for organics in the US as a consequence of worsening environmental and social outcomes, including the 'continued demise of [the] family farm'.[39] Bob met Terry Gips from the US, Nic Lampkin of Aberystwyth, and Bernward Geier, a research scientist working with Hardy who was 'very impressive'.[40] The experience of being among leaders in the international organic movement was nourishing, validating and invaluable, but Bob observed, 'It is still a lonely road for many professional people when they profess a philosophy that questions the very basis of modern science and technology.'[41] Bob himself made an impact on Holger Kahl, who would later become a significant feature in Bob's professional life. Holger made a note about Bob in his journal: 'Who is the peculiar man with leather belt and pouch, just about the only conference participant in shorts, bursting with energy and one of the keenest when we cue [sic] up for the organic wine tasting?! He's from New Zealand I am told, works at Lincoln College and has established an organic demonstration area.'[42]

The benefit to New Zealand's organic movement in having Bob attend the IFOAM conference (not to mention the rest of his research trip, which took in Britain and North America) is inestimable. *Soil & Health* readers were suddenly introduced to a broader view of the work their organisation had been doing since the 1940s, as well as new research and exemplary practice. For example, an article by Hardy was included in the Autumn 1985 *Soil & Health*, in which he pointed out the economic problems inherent in conventional farming and argued for a 'sustainable agriculture', because 'only ecologically-sound solutions will also be economically viable'.[43] Bernward Geier updated readers on progress the following year.[44]

IFOAM, which by 1984 had over 100 members representing 50 countries, released an international organic standard that year, and during 1984 and 1985 the New Zealand Biological Producers Council (NZBPC) used this for inspections of organic properties for certification.[45] Chris and Jenny May, working with others, also developed a New Zealand standard, drawing heavily on the IFOAM one and the British Soil Association's standard. By mid-1985 a BioGro logo had been developed and a statement that produce bearing the logo had been grown 'in accordance with the New Zealand Biological Producers Council's standards' appeared on the cover of *Soil & Health*.[46] By the end of 1985, around 50 producers had been certified.[47]

•

In 1985 Perry Spiller joined the national organic scene. Working together, Perry, Chris May and Bob would reshape the nature of the organic farming movement in New Zealand.

Perry had joined the Soil Association in 1968 and been vice president since 1980.[48] In 1984 he was elected onto the NZBPC, and in 1985 he was contracted through Project GRO to work on the advisory service that Chris May had been instrumental in establishing.[49] As education officer, Perry took on a central leadership role within the Soil Association and the organic movement generally. This was especially the case when, in 1985, Chris and Jenny May relocated to Christchurch in order for Chris to undertake postgraduate studies at Lincoln.[50] In one month alone Perry represented the association at the Environmental Conservation Organisation (ECO), the parliamentary select committee on Plant Variety Rights legislation and the inaugural meeting of the New Zealand Pesticides Action Network (PAN), and was on the leadership team for NZBPC.[51] When Chris became NZBPC president in 1986, a position he held until March 1988, Perry took over as president of the Soil Association.[52] Bob became vice president of the NZBPC, also in 1986.

Throughout this period Bob began to agitate for change in government policy, writing a stream of letters to government ministers and other members of parliament. It was a path that was unlikely to end well: writing to an MP as a private citizen is one thing, but writing as an employee of a government-funded organisation to complain about the lack of government support for one's own area of interest was possibly reckless. His capacity for bullishness can be seen in his letters to the Minister of Agriculture, Colin Moyle. On 12 February 1985, Bob wrote to Moyle to inform him that he had returned from four months' sabbatical leave, during which he had attended the IFOAM 1984 conference and visited a host of other locations of interest. He enclosed various reports that he thought might be of interest to Moyle.[53] Moyle's response was somewhat tetchy: 'I have briefly perused the literature you forwarded and note the "Update: Biological Husbandry in New Zealand" … which you presumably wrote.' Moyle then shared some updates on organic research being undertaken through his ministry.[54]

Bob replied in June: 'I feel that the concepts which we represent are not getting across to you.' He was 'well aware' of the research Moyle had referred to and added that he and others 'are urging you to do more than "briefly peruse the literature" available on the viability of sustainable agriculture and to take the initiative in urging government to look closely at the inter-relationships between agriculture, health, environment, tourism, and the exports of our primary produce.'[55]

Moyle's reply was terse: 'I make no apologies for the fact that personally I have to limit the time which I can devote to studying the details of everything supplied by every enthusiastic advocate.' He warned: 'I believe that the statement in your most recent letter to me that the concepts you represent are not getting across to me is incorrect and ill advised.' He spoke of what he considered 'significant initiatives being made by government-funded research, even as also is the contribution which government is making to your work through its funding of Lincoln College'. If Bob's work was as valuable as he claimed it to be, it would 'not be overlooked by your departmental head or by the College Principal'. Moyle concluded by recommending that Bob concentrate instead 'on achieving full support at that level'.[56] Not to be put off, Bob wrote again the following month, asking for the minister's financial assistance in getting him to IFOAM.[57] Moyle refused.[58]

Bob also approached the Minister of Tourism, Mike Moore, and wrote to Simon Upton, an Opposition MP, saying he did not think Colin Moyle could 'conceive of the importance of the movement or the urgent need'. He added that his research unit at Lincoln had the capacity to be the equivalent of Rodale, Witzenhausen, Oberwil or Santa Cruz, 'but we must have pressure to show the Lincoln College Council that such a unit is *respectable*'. Upton replied saying he would suggest to his colleague, Ruth Richardson, that they should call on Bob to discuss the matter.[59] It is easy to see how communications such as these could become problematic if relayed to the wrong people.

Nor should it be imagined that relations within the organic movement were trouble-free. In October 1985, 18 people representing the NZBPC, the Biodynamic Farming and Gardening Association, the Soil Association and the Doubleday Research Association of New Zealand met for three days to focus on 'the future and structure' of the NZBPC, and to map out 'strategy and policy for the future'.[60] Such collaboration was world-leading, but it was perhaps also tenuous. In 1986 rule changes to the NZBPC resulted in what one member of the Biodynamic Farming and Gardening Association viewed as 'a near total takeover of the Council's Executive Committee by Soil Association people', a 'coup' done 'quite ruthlessly' and 'with a certain amount of relish. I find the whole scene deplorable.'[61] The Biodynamic Farming and Gardening Association confirmed that it was 'disappointed and concerned' by the way in which Soil Association members filled 'important positions on the Executive Council through the Delegate representation'; 'our council remains firm and resolved' in giving full support to the 'original impulse of the NZBPC based on a foundation of co-operative understanding'.[62]

Bob led conversations within the NZBPC in 1986, about the need to professionalise BioGro organic certification inspections and move away from a volunteer-led model. This move was partly prompted by discussions within the Canterbury Organic Producers group. Tony Mallard, who led the group, acknowledged that 'it depended on Bob Crowder to a great extent; people [are] more than capable to accompany Bob but feel [they] need his professional touch as the leader'.[63] Bob argued for the inspectorate to become 'a fully qualified body of people ... official inspectors need to have recognised qualifications from our establishments'. The NZBPC was 'fortunate to have on our Executive at least four who have very good qualifications even in the eyes of the scientific community', which was 'important when setting up an inspectorate if [we] have to answer to the Ministry of Agriculture'. If BioGro inspectors were qualified to the same standard as MAF officers then 'obviously we have a very good negotiating position to have our standards and inspectors respected'. Bob proposed that he, Bruce Collins, Chris May and Jonathan Toye would 'try to lay down the basic foundations for a course structure for the first inspectorate'.[64]

In October 1986 Bob outlined the resulting basis for the BioGro inspection training system to the Canterbury Organic Producers. A three-day course for inspectors would cover IFOAM regulations, entomology training, soil characteristics, communication skills and farm case studies.[65] Later in the year Bob pushed for further work on this professionalisation of the inspectorate; his ideas were accepted by the committee.[66]

There were some challenges to the NZBPC certification processes in 1986. In one example, a difference of opinion regarding the spraying of a kiwifruit crop to meet export requirements left the grower furious: 'I have had a "gutsful" of the NZBPC which, unless it gets its act together and operates in a professional manner, will win little respect in any quarter,' he wrote.[67] Bob tended to be pragmatic in such instances. Minutes of the NZBPC executive meeting record his view that since BioGro needed more experiences with export, he would agree to certify this grower. However, it would be done on the 'strict understanding that we are taking this action as a purely experimental marketing experience and not as a precedent ...'[68]

But some horticulturalists felt the system for becoming certified organic was too onerous – in particular the three-year conversion process where growers could not claim any organic status for at least a year, despite the likelihood of some loss of yield from the sudden withdrawal of chemical inputs. Later in 1986 a new organisation, the NZ Organic Suppliers Co., announced a plan to launch a standard with a much simpler process of certification for what it called 'conservation'-grade produce. The grade was 'not intended to compete or detract from Bio-Gro'; the company was merely 'fulfilling the market demand for simple "residue free" produce', saying 'such a move was inevitable in NZ as it has been overseas'.[69] Needless to say, BioGro was opposed to the idea.[70] The conversation about different grades reflected gripes within the sector about the perceived 'pickiness' of the NZBPC, but this didn't dampen the spirits of those involved. 'It was exciting, all new,' said Bob. 'It is a question of being stimulated towards a philosophy, a belief ... it was an advance.'[71]

Working as volunteers, Bob was responsible for inspecting organic properties for BioGro certification in the South Island and Peter Proctor for properties in the North. They would then come together for weekend-long meetings with Chris to decide whether particular farming activities were acceptable or not. These meetings usually took place in Auckland and the debates were sometimes fiery. 'We had arguments,' Bob confessed frankly.[72] The BioGro standard, as the New Zealand standard came to be known, was released in 1986.

If the NZBPC was picky, it was also energetic in seeking solutions to the difficulties faced by its producers and in rising to challenges from MAF. MAF was obviously very interested in what the NZBPC was doing, as demonstrated in a letter from senior MAF official Robin Scott to the NZBPC in July 1986. He said he had seen how rock phosphates could be used instead of superphosphate on 'about half of New Zealand's soils': 'I believe it is now possible to explore the prospects of biological farming systems more fully in respect to the production of biologically grown meat products.' He wanted to meet with the NZBPC to discuss areas in the standards 'which I believe need clarification and possibly justification' and emphasised that he wanted to focus on meat, where he saw strong potential, and not horticulture or grain, which he believed would be more difficult to grow organically.[73]

In August 1986 a team consisting of Chris, Perry and Marinus La Rooj met with seven members of MAF in Wellington to discuss the issues raised by Robin Scott and to voice their own concerns. Marinus remarked, 'For the first time I think they actually listened to us, instead of just hearing what we had to say.'[74] On the same trip, Chris May met with MAF's Chief Advisory Officer (Plant Exports) 'to discuss the difficulties being experienced by BioGro producers in attempting the export of certified kiwifruit'.[75]

At the end of 1986, the earlier clash with the Biodynamic Farming and Gardening Association came to the fore again. President of the Soil Association Perry Spiller wrote to the NZBPC to say the national council 'totally opposes joint promotion of BioGro with any other standard'. The discussion within the NZBPC on this point was robust, and Bob made his views plain:

> My understanding of NZBPC originally was that everybody agreed that we needed to have a co-ordinated effort to get the biological issues of the world to the forefront; my concern [is that] we have got to be credible to as many people as we possibly can ... [the] starting point of getting [the] message across to people

is the promotion of fundamental agriculture which is as close to conventional agriculture as it can be without violating [the organic] standards. Increasing numbers of people are brought into organics in this way; BPC [is] dealing with these people and I personally believe it is very detrimental to our work, and for my work, to be in an organisation that is working with biodynamics … the fact is that many farmers and many scientists can go along with the practices of good husbandry but not as soon as moving into biodynamic philosophy.[76]

The results of the discussion were not conclusive, but Bob's concern for his reputation and the credibility of the movement that he was aligning himself with (and actively cultivating) was palpable and, again, pragmatic.

Around this time Bob led a workshop on organics in Ashburton, which Brian Patchett attended with his employee, agronomist David Musgrave. David had purchased his family farm, Waihi Bush, in 1982 and was interested in Bob's approach. By 1988 he had converted Waihi Bush to organics. The Musgraves hosted a field day for COP in 1988.[77] As auditor, Bob was heavily involved in the conversion and recognised in David the qualities needed to move BioGro forwards. He convinced David to pick up audit work for BioGro, and David was soon rewriting parts of the standards and developing new standards for organic manufacturing and honey production. In July 1990 David gave a presentation on the standards at a COP event at Lincoln.[78] He and his co-worker at Dalgety's, Jon Manhire, would go on to play significant roles in Bob's organic legacy at Lincoln.

•

The 1986 IFOAM conference was held at the Santa Cruz campus of the University of California. While there, Bob visited the Californian Certified Organic Foods group, which he described as an 'enthusiastic band of qualified workers fighting hard on a restricted budget to make philosophy work under difficulties'.[79] He could have been describing the NZBPC. Meanwhile, the NZBPC had given Bob authorisation to represent it and to vote at the IFOAM General Assembly meeting – an important moment in his development.[80] At Santa Cruz Bob put on an even bigger display of produce than he had at the 1984 conference.[81] It was a tremendous effort, especially since, as Perry remarked, Bob's attendance came mostly out of his own pocket: financial support from Lincoln College had amounted to less than a quarter of his costs.[82]

The Santa Cruz conference was notable for several reasons, as Bob later reported. First, Bernward Geier was appointed secretary-general of IFOAM and would leave his role at Witzenhausen to take up the position from February 1987.

Bob photographed at the 1986 IFOAM conference, Santa Cruz. (Photo: Robert Beutleman)

Second, pressure was growing for IFOAM to regulate the global trade in organics: to 'ultimately block trade in any organic produce other than that grown to IFOAM approved standards'.[83] There were obvious implications for the NZBPC: it would present 'an urgent and interesting scenario for organic producers interested in export particularly to Europe'.[84] Regionalisation of IFOAM was another issue, but in the end this was not agreed on due to a lack of confidence from IFOAM that regional groups had enough expertise to monitor their own standards. 'It is essential, therefore, that over the next two years the whole administrative procedure of the organic movement in New Zealand, and in co-operation with Australia, should be streamlined and professionalised,' Bob declared.[85]

Federated Farmers had played an important role in bringing the nascent organic sector together in 1982 at Flock House. Despite strong reservations from some quarters about Federated Farmers' views on organics, the 1987 Soil Association conference (the organisation changed its name that year to the Soil & Health Association of New Zealand) demonstrated that 'The Feds' were still engaged. Chamberlain, senior vice president of Federated Farmers, opened the conference, remarking encouragingly 'that the members of your association have an important role to play in New Zealand agriculture'. He explained that he had become interested in biological farming as a result of his interactions with Peter Waugh, who had worked hard to get biological farming on the Federated Farmers agenda. Brian Chamberlain's address focused on economic policies, technological developments and marketing strategies, and was reported in *Soil & Health*.[86]

In the same issue, national science programme manager at MAF, Dr Robin Scott, wrote about evolving policy and the prospects for organics. Clearly, this weaving together of Federated Farmers and the ministry was part of a strategic approach by the Soil Association to mainstream its work. Indeed, Robin Scott was writing about a decision just made by the ministry to assist in developing and promoting organically grown food. He acknowledged problems with the 1983 report that compared conventional and organic farms and noted that a more comprehensive study was currently being undertaken.[87]

Bob had spoken with Hardy Vogtmann about the inherently flawed approach of comparative studies of this kind. Hardy replied:

> I feel we are losing a lot of resources and time if we start comparing. I know that if I have a trial plot that it will grow if I put on a lot of nitrogen fertilizer and that it won't if I don't. But I can only answer the question when I get all the elements of the system working. That means that I have to work in a complete system. And then, if you start comparing, you have to compare complete systems and from my knowledge of comparative trials throughout the world there is none that

compares systems, they usually only compare sub-systems or even elements of a system. So it is pointless to start comparing systems, it costs a lot of money and won't be dynamic.[88]

Robin Scott outlined MAF's strategy: to carry out market research to establish overseas export demand; to establish an acceptable framework for organic certification; and to attend to 'the development of the ways and means of producing the product'.[89] The second step in this process – the framework for certification – is important. As Robin said, 'The use of international trademarks, use of the Biological Producers Council or other such trademarks, or creation of a MAF organic standard are all options to be evaluated.'[90] That MAF was considering developing its own standard in 1987 is certainly of significance.

Overall, Robin's approach, like that of Brian Chamberlain, was encouraging: 'I look forward to the day New Zealand is a major international supplier of organically grown food. Getting there will be an exciting challenge in which the ministry wishes to participate along with organisations such as the Biological Producers Council.'[91]

•

Around the middle of 1987 the results of the second major study of organic farms were released, authored by Anthony Haystead of MAFTech (MAF's research arm). The study was, again, a simple comparison between organic and conventional farms in different areas. Haystead acknowledged that the 1983 report 'was not well received by the organic movement in New Zealand since it offered little support for its aspirations and endeavours'.[92] This second study, which had been triggered by the outcomes of the previous one, was meant to last for six years but had been halved, ostensibly due to the commercialisation of MAF services (an innovation of the fourth Labour Government). The organic community suspected the study was halted because organics was irrefutably coming out on top.[93]

The results of the project were therefore to be treated with some caution, though Haystead believed they were 'still significant' and had 'contributed in no small way to the change in MAF policy on organic farming'.[94] If nothing else, the study had shown that organic farms could hold their own, which was slightly more encouraging than the 1983 report. For example, the plant analyses showed 'no significant differences between organic and conventional farms … [and there was] no indication that organic methods lead to a degeneration of pastures and a loss of clover from the sward'. His comments on stock losses typify the extremely cautious optimism of the report:

Break time at the BHU. Bob with Brendan Hoare and visitors (and Jill the dog), c.1987.

The major stock losses in the study occurred on the conventional farms and, if anything, animals were less healthy under conventional management. From such a small study it would be foolish to conclude that this will always be the case. However, it is reasonable to conclude that a change to organic methods is not necessarily accompanied by serious animal health problems.'[95]

In November 1987 a discussion document was presented to the NZBPC by MAF, who sought to bring the two parties closer. MAF staff member and NZBPC committee member Jo Springett described the situation: 'MAF felt the need for NZ producers to get into organic production was urgent, as increasing consumer awareness and concern over chemicals made selling conventional produce increasingly difficult.'[96] The NZBPC, on the other hand, understood this as a MAF attempt to take over BioGro.[97] Indeed, MAF had even proposed a new name for the NZBPC: the Organic Food Standards Board. Chris May, Perry Spiller and Bob were delegated to meet with MAF and take the conversation forward.[98] The outcomes of this meeting are not recorded. However, the arrangement the three came to with MAF led to MAF's financial support in bringing the IFOAM team to New Zealand in 1988 to undertake the BioGro accreditation. For now, at least, the tussle with MAF seemed to have been put to rest.

In 1986 Bob's organic unit at Lincoln was variously called the 'Biological Husbandry Demonstration Unit' and sometimes the 'Biological Husbandry Research Unit'.[99] This slight shift in naming reflected Bob's aspirations to move from demonstration to research, and with Geoff Barnett now employed as a technician this made perfect sense. The simpler 'Biological Husbandry Unit' began to come into use in 1986 and would later be shortened to the 'BHU'.[100]

The development of semi-intensive beds in 1985 and agricultural rotation in 1987 also reflected the maturing vision for the unit.[101] The first was a six-year crop rotation 'involving the semi-intensive production of mainly vegetable production'.[102] It received no inputs 'other than rotational manipulation with fertility building breaks and green crop incorporation'.[103] The second was developed when Mike Daly of MAF joined the unit to undertake research on wheat production in rotation with amaranthus and dried beans. The unit's agricultural rotation, also called the extensive rotation, was built on Canterbury's traditional system of three years of grass and legumes (normally clover) grazed by livestock followed by three years of cropping. At Lincoln, Bob adjusted the old system by incorporating a mixed herb ley pasture and managed the rotation without animals.[104]

Enabling these developments, as Bob pointed out, were 'the declining fortunes of conventional research investment'. This meant that large sections of the Horticultural Research Area were no longer in use.[105] With Geoff's able management, Bob was able to incorporate these pieces of land into the steadily growing organics unit. By the beginning of 1987 the unit had expanded to more than three hectares and included managed shelter belts; the original intensive garden; two plastic tunnel houses; a rotation demonstration covering just under 0.4 hectares that included work on alternative strategies for weed control; an area of managed mixed herb ley, green crops and amaranthus; and a mixed orchard that now covered close to 0.4 hectares.[106] By 1988 the unit covered six hectares.[107] Further expansion was just around the corner.

However, changes in tax laws introduced by the fourth Labour Government were affecting Project GRO and confounding the research strategy: 'Rogernomics' (the name given to the neo-liberal economics introduced into New Zealand by the Minister of Finance, Roger Douglas) 'has clawed its way in here, too. Any contributions to Lincoln College, after the expiry of the current contract in 1989, are subject to 10 percent GST. So it seems that what is tax deductible one way, is taxable the other.' As a result of this coming change, Soil & Health asked Lincoln to 'accept the full cost of maintaining a technician in the Biological Husbandry Demonstration Unit … continued support would be better directed at subsidising a second technician to assist Bob Crowder.' It sought 'to gain further commitment from the College, rather than just leave the status quo unchanged'.[108]

In 1987 Geoff left Lincoln to travel and Bob offered the technician's position – still largely funded through Soil & Health's Project GRO – to Brendan Hoare. Brendan set to work managing all aspects of the unit: the intensive beds, the orchard, the tunnel houses, and 'beefing up the whole biodiversity of the site', especially by developing the mixed shelters. Brendan involved others in the site and managed the labourers from the Wwoof scheme – Weekend (later 'Willing') Workers on Organic Farms. The labourers, known as 'Wwoofers', sometimes pitched their tents in the research area and stayed for several weeks at a time. According to Brendan the area was 'full of young people who were studious and had travelled the world, and he [Bob] really looked after them'.[109]

The Lincoln unit was a very different place from the windswept empty paddocks Bob arrived at some 20 years earlier and had developed something of a cult status within the organic movement.

Quinoa and sunflowers included in the semi intensive crop rotation.

6: New Zealand organics on the world stage, 1988–92

IN MARCH 1988 Bob took over the role of president of the NZBPC. Under his leadership the organisation improved its professionalism and established a technical committee to consider the technicalities of defining what was and what was not allowed in organic farming systems, and under what conditions.[1] New potential inputs and practices were changing rapidly and the committee provided relatively quick responses to unanticipated queries. The NZBPC was now managing at least 126 organic licences (i.e. certifications) and had improved its inspection systems.[2] This professionalisation was critical, as Bob had earlier pointed out, because it would determine BioGro's position within IFOAM as the world body developed its international accreditation programme. Inspector training courses, which Bob developed, brought staff from MAF Qual. and MAF Tech. together with non-MAF inspectors: 'One could not have envisaged such a meeting even a few years ago,' Bob wrote in 1989.[3]

Bob's efforts to professionalise BioGro put New Zealand well ahead of Australia in terms of developing commercial organic infrastructure. Bob had joined the National Association for Sustainable Agriculture Australia (NASAA) in mid-1987; Australia, he now said, was 'a long way behind in certification procedures'.[4] Secretary of NASAA Sandy Fritz had written to Bob at the time to update him on the question being discussed within IFOAM about devolving decision-making on organic certifications to regional bodies, rather than a single, global body, something she had been discussing in correspondence with IFOAM's secretary-general, Bernward Geier. Sandy envisaged 'a structure or rules ... that would allow for maximum independence of regional activities along with as much input from regions into international discussions and activities as possible – while allowing for efficiency'. The question of regionalisation would still be on the table as late as 1997. Sandy also sought Bob's knowledge of non-chemical controls for apple dimpling bug and plague thrips for the Advisory Committee to the Minister for Agriculture, Lands and Forests on Agricultural and Allied Chemicals, on which she sat. Meanwhile, she said, she was teaching a new paper in agroecology and sustainable systems at the Australian National University, using Miguel Alteiri's book as the text.[5] Exchanges like this illustrate how useful the international system that had developed within IFOAM could be, and of course emphasise Bob's own role in influencing the direction of organics internationally.

The increased professionalisation of the NZBPC gave the government confidence to support the accreditation process. IFOAM staff (Bernward Geier and Jan von Ledebur) would travel to New Zealand to assess the BioGro standards and procedures, which would involve meeting with key personnel from the NZBPC as well as organic farmers and staff from MAF. But the trip would also serve the purpose of bringing the Pacific region more firmly into the awareness of IFOAM's still somewhat Eurocentric thinking.[6] Bernward's trip to New Zealand was funded through a grant from the Market Development Board, which had been constituted by the government in 1986 to 'foster New Zealand's foreign exchange earnings by promoting, encouraging, financing and assisting activities carried out or proposed to be carried out by a sector group or combination of sector groups for the purpose of developing overseas markets for New Zealand goods and services'.[7] Additional support was provided by the Biodynamic Farming and Gardening Association, Soil & Health Association and NZBPC; MAF covered Jan's costs.[8]

Bob wrote to Sandy in February 1988 to confirm that Bernward and Jan would be visiting New Zealand in May and invited a representative from NASAA to join the discussions, 'in order that you will be able to see what is involved in such a national inspection'. He expressed the hope that MAF would contribute towards the cost.[9] Sandy was enthusiastic: 'Australia is interested in inclusion in the [IFOAM] International Directory and we would be most interested to observe what is involved and to learn from you folks who have been at it longer.'[10] MAF did in fact part-fund Australia's participation.

By April the tour of inspection was organised. It would begin with a visit to the BioGro office in Auckland, followed by visits to the Bay of Islands, Northland and Rotorua. MAF would take the IFOAM visitors to Ruakura, Flock House in Palmerston North and Wallaceville in Upper Hutt. Discussions were planned between MAF, NZBPC and IFOAM in Wellington, then the party would move on to Christchurch to visit MAF and finish with the Soil & Health conference at Lincoln. One of the most important topics for discussion was the fierce disagreement between MAF and the NZBPC (and IFOAM) about the use of anthelmintic treatments for internal parasites in sheep. But the problem was resolved by a compromise that allowed sheep to be treated provided they were not sold as organic within a certain withholding period.[11] For this occasion, most uncharacteristically – and at Perry's insistence – Bob wore a tie.[12]

After the tour, it was time for the IFOAM team to examine the technical details relating to New Zealand's approach to organic certification. This afforded an opportunity to bring Australia into the Technical Committee of IFOAM, which would make the Australian assessment process much simpler should they pursue

Coppiced poplars for energy at the BHU – part of Bob's holistic vision.

it.[13] As a 'corresponding member' of the Technical Committee (since earlier in the year), Bob's influence in this was apparent.

Sandy wrote to Perry Spiller after the trip to express her thanks for the ways in which 'the New Zealanders I met went out of their way so much to make us (Jan, Bernward and myself) comfortable'. Contacts she had made would be invaluable in developing the Australian organic export market: 'Making overseas contacts now could be crucial in NASAA getting some funds for its work.'[14]

The assessment road show of 1988 probably marks the beginning of an Australasian organic sector. Although there was some discussion between NASAA and BioGro about a joint logo between the two organisations, this did not eventuate.[15] Shared branding was topical: despite small flare-ups in the past between the Biodynamic Farming and Gardening Association and the NZBPC, the two had now produced a joint poster. Jan von Ledebur and Bernward Geier expressed surprise at this; such collaboration was unheard of in Europe. As Perry Spiller recalled, 'They commandeered as many as they could to take back to Europe to show Europeans how NZ organics and biodynamics were global leaders in mutual co-operation.'[16]

Despite the apparent success of the IFOAM visit, NZBPC minutes record a drawn-out process afterwards that became increasingly frustrating. By early 1989 there was still no confirmation that BioGro would be included in the IFOAM Register of Internationally Approved Organic Standards.[17] At the IFOAM conference in Ouagadougou, Burkina Faso, early that year, Bob had been elected to the IFOAM World Board.[18] In this capacity, he explained to NZBPC members, he would be able to 'ease' the BioGro standards 'through to their acceptance'.[19]

His report proved controversial. The first hint of trouble followed a July letter from a New Zealand distributor to a German importer, Naturkost, informing them that the NZBPC had 'met with and have had their "Bio gro" label certified by IFOAM'.[20] Naturkost promptly forwarded the letter to IFOAM. Bernward wrote to Bob in August to ask how such a statement could be made, 'since IFOAM always stressed that they will not go public with results of the evaluation of single associations, but rather wait [until] they have a "package" of evaluated organizations together. I do know,' he added, 'about the confusion that was around the evaluation of BPC based on the information that you got after the Ougadougou General Assembly … July was a bit early to make reference to the IFOAM evaluation, especially since the contract has not been signed.'[21]

Bob apologised for the 'premature article' and replied that the episode 'underlines the need to get this business sorted out quickly so we can use what has been paid for': the confusion was as a result of '*NO action* leading to frustration'. The IFOAM Technical Committee had not yet made any statement on the evaluation carried out: 'Where do we stand?' Bob asked.[22] Certainly, the situation was confusing; as Jan von Ledebur had explained to Perry Spiller in July, the system had changed since the evaluation was completed:

> After thorough and extensive discussion, the IFOAM-TC and board of directors have decided to build the future system of evaluation and publication of reports on a contract system to ensure just and fair use of the information contained in the reports. This contract clarifies the conditions of how to use the evaluation reports, it guarantees a system of regular actualization of the information in the reports and it also … is concerned with the question of how to finance this work of IFOAM.[23]

IFOAM had sent a contract to NZBPC to sign and was awaiting its return. But, Jan added – and this must have irked the New Zealand crew – they would receive 'the whole bundle of reports, after the respective contracts were signed by the organizations concerned'.[24] This implied, as Bernward later clarified, that nothing was official until a number of contracts with certifying bodies from around the world had been completed. The NZBPC signed the contracts as written and the reports were finally received by February 1990.²⁵ Bob acknowledged that the process 'is not all we could have wished for'; it meant that organisations who wanted to know whether BioGro met organic standards could purchase the evaluation report and make up their own minds; IFOAM would not pass a definitive judgement.²⁶

Exactly what came of the IFOAM trip to New Zealand is not clear from the record and may reflect the ongoing evolution of IFOAM's programme of evaluating certification bodies who were IFOAM members, which had commenced in 1985. Its intention was for those members to gain 'experience regarding recognition among certifiers', a reflection of the proliferation of organic markets internationally.²⁷ IFOAM's accreditation programme was only approved in 1990 and work commenced on it in 1992. In 1995 the first accreditations of certification bodies were made. NASAA was one of the first to be accredited in this programme.²⁸

Regardless of the evaluation outcome, a number of positives emerged from this process. Bob's role in developing networks with Australia was critical and, as he summarised later, the 'main contribution that I gave to IFOAM was to bring New Zealand and Australia into the framework of thinking ... to draw attention to the fact that there were other places outside of Europe that had an ... organic movement and [were] working towards standards.'²⁹

In September 1989 he attended a special executive meeting of NASAA. His confidential report to the NZBPC on the meeting noted that NASAA had suffered from a shortage of money and 'some *poor inspections* by *poor inspectors* lacking in *qualifications* and *credibility* for the rapidly rising tide of interest by "*professional*" organisations'. The volunteer inspectors had been replaced by paid inspectors, a move that had caused stress and increased fees. Bob suggested to NASAA that past inspectors be allowed to reapply for their positions, provided they were willing to receive extra training. The proposal was accepted.³⁰ Overall, he thought, the NZBPC had more competent administration; it was more democratic and its standards and inspections were better.³¹ BioGro could help the Australians to up their game.

The New Zealand tour of 1988 also helped to unite the domestic organic sector. It had concluded with the Soil & Health conference at Lincoln: 'NZ's Growing Future'. Six hundred people attended, which made it the largest organics conference to date in New Zealand.³² The conference opened with an address by Lincoln College's principal, Professor Bruce Ross, who announced that the BHU would be fully funded, its size would increase to 10 hectares, and Bob would be appointed as the unit's director. Bob wrote afterwards: 'The Biological Husbandry Unit has come of age and been accepted by Lincoln College as an integral part of the Horticultural Research Unit.'³³ At the festivities on the Saturday night, 'a certain senior lecturer was prominent amongst the [Morris] dancers, arrayed in bells, braces and top hat, plus a grin that reached from ear to ear.'³⁴

Bruce Ross also spoke of 'the advantages to New Zealand of pursuing organic methods of food production'. This theme was echoed by Robin Scott of MAF, who 'told of new directions in organic growing and said he hoped the majority of farms in New Zealand would one day be organic'.[35]

Bernward Geier was a principal drawcard at the event and spoke of the rapid increase in interest in organic farming but the relatively low numbers of converted farms. He recommended focusing on advice to farmers, organic education programmes (including for scientists) and a change in government rationale. To MAF he said, 'Please keep in mind that organic farming is not only economics and I think in the ministry that message has to be spread a little more. It is not only exports, it's your country that will be in a lot better shape if you convert your farmers.'[36] The organic farming movement, he observed, posed a threat 'to the people who make a lot of profit out of this conventional farming system'.[37]

The conference was Bob's chance to shine at his own institution. Visitors were able to visit the BHU and see research projects in action. These included using flowering parsnips to control leaf roller and codlin moth in the orchards, the economics of a coppicing shelter belt, cereals like amaranthus in a six-year rotation, and of course the maturing compost heaps.[38]

Brendan Hoare's time working with Bob at the BHU – as it was by then generally called – came to an end in 1989 when he was 24:

> I got to practically apply what was in my head … I remember feeling really honoured … I had a diploma, a degree, and had managed the BHU for two and a half years … I was always interested in the wider political sphere as well, which was always great with Bob because … the conversations were expansive. And when we met with growers it was never just about [growing crops] … It was always about influencing a bigger, wider framework.[39]

Brendan recalled that, as a teacher and employer, Bob 'was enabling and encouraging, he had his own opinions but was just as quick to laugh at them and learn from others as well. He was also a humble learner.' Having supporters on hand was necessary: '[Bob] was never treated well, and [in] trying so hard to do so much good work, got sort of brutalised.'[40]

Meanwhile, conversations were continuing with MAF, and it seems likely that Bob's impatience with the IFOAM evaluation process was partly prompted by a desire to outflank MAF in its organic aspirations. Kim Stevenson had taken over as MAF's national organics manager, and in February 1989 he attended an NZBPC meeting to discuss, in confidence, some of the resourcing MAF was preparing to provide to the sector. He gave assurance that MAF 'was not trying to absorb BPC' and outlined a plan to conduct a market survey to establish what the market wanted in terms of organic production.[41]

The earlier concern among the NZBPC leadership regarding MAF's role in organics took a new turn in the second half of 1989. Stevenson wrote to Bernward Geier to say that during the year, 'negotiations have been proceeding with the Biological Producers Council regarding the MAF taking a role in certification and inspection of farms'.[42] He planned to attend the IFOAM working group on certification in October. Kim Stevenson sought key contacts 'in various countries so I may go and talk with certification authorities and technical research people about what they are doing and so we may have an open exchange of ideas'. He was also interested to learn more about the IFOAM rules around transition periods for new farms entering into organics, saying, 'We need some kind of understanding so we can encourage as many farmers as possible to become organic without having unnecessarily restrictive transition periods.' Furthermore, he 'would also like to explore the possibilities of the New Zealand Ministry of Agriculture and Fisheries becoming an independently IFOAM-ratified organic certifying authority, as there are some commercial problems with becoming too closely associated with the commercial label of the BPC'.[43]

Bernward's reply is not extant, but Kim's response to it is. It is apparent that Bernward encouraged Kim to work closely with the NZBPC and Bob. 'We have been trying to work as much as possible with the NZ Biological Producers Council (of which Bob Crowder is president) but have unfortunately found them to be slightly impractical,' Kim complained.[44] 'We are very concerned about the restrictive nature of the BPC transition period':

> Many NZ farmers are in a very tight financial position due to a number of causes, and while many have an interest in moving towards organic systems, the current requirement for a 'zero' year with not even partial registration means that this is financially impossible (as yields fall but there is no chance of a price premium). Thus while we are getting a high level of interest to our media coverage of organics, the area registered organic in NZ is essentially static or maybe even declining. The MAF does not believe that this is a good way to get an adoption and expansion of organic principles.
>
> Thus, the NZ Ministry of Agriculture & Fisheries wishes to set itself up as an independent certifying authority for organics and instigate a 3-year transition period: provided residue levels are not too high, then the first year without chemicals would become Transition Organic Year 1; then next year, Year 2; Year 3; and then full organic.
>
> We would use exactly the same NZ standards as presented to you by Bob Crowder – the BPC do not own these standards as they may have indicated to you.[45]

Such statements must have been surprising to IFOAM, given their experience in New Zealand the previous year, and were likely to threaten the still unresolved issue of IFOAM accreditation. Bernward forwarded the message to Bob: 'Dear Bob, I think you should know [of] this letter and maybe comment on how I should respond.'[46] Bob replied, 'You were quite right to refer it to us here in NZ to deal with … I think Kim would like to take us over, perhaps it would be a good idea but not this way … Be careful with our MAF please.'[47] (In fact, on this last point, Bob even commented to Bernward the following month that he thought the NZBPC should simply '*sell* BioGro to MAF and use the money to establish a *well-financed pressure group* to continue to act as the *conscience* of the organic movement'.[48])

Bob took the matter to Ian Cornforth at MAF:

> During the organic standards meeting between the BPC, MAF and IFOAM in May 1988 … the Biological Producers Council was given the assurance that MAF was not about to develop an alternative certification procedure.
>
> … It now appears that policy has changed and the MAF is seeking to negotiate with IFOAM with no prior discussion with our own organisation.
>
> To date all debate and controversy has been kept as an internal matter and has been amicably settled through discussion. The result is an excellent image in IFOAM of a country where government and the organic organisations are working in close co-operation towards a single goal.
>
> The letters by Kim Stevenson, on behalf of MAF, to IFOAM came as a shock to myself during my recent attendance at the IFOAM board meeting in Germany. The recent letter from Kim Stevenson forwarded to me from IFOAM came as an even greater surprise.[49]

Bob addressed a number of important points, including the transition phase, barriers to export and ownership of the BioGro standards. His comment on the last of these was couched in curiously ambiguous terms:

> It is … regrettable that there is still confusion over who owns what with regard to the organic standards. It is the Bio-Gro trademark that is owned by the Biological Producers Council of New Zealand which is based on the organic standards drawn up by the Biological Producers Council. Although Jo Springett was heavily involved with those standards it was as an elected member of the Executive Committee of the Biological Producers Council that she carried out the task. We fully acknowledge that she was able to do this work for us while working as a MAF officer in the same way that I am indebted to Lincoln College for the same reason.
>
> In any event all organic standards are based on the fundamentals laid down by IFOAM and the New Zealand standards were developed with a heavy reliance on those of the Soil Association of Great Britain and the Californian Standards.[50]

So strained was the relationship that Bob even secured funding for Els Wynen of NASAA to attend IFOAM conferences in Linz on behalf of NASAA and the NZBPC, mainly 'to have an observer there in order to get some independent feedback on what Kim Stevenson is doing … We need to know how he presents MAF compared to the BPC.'[51] MAF's response to Bob's request for a high-level meeting is not available, but for the time being MAF seemed to back away from the idea of establishing a competing agency to the NZBPC.

Bob continued to present a collaborative image of the NZBPC and MAF and perhaps stretched the truth slightly in saying they were working hard to 'develop the MAF/BPC relationship'. By early 1990 it was gratifying to note that 'several MAF personnel are now actively entering the BPC Inspectorate'. This was a slow process, however: 'Negotiations with MAF have not eventuated to the extent envisaged this time last year,' Bob lamented in his annual president's report.[52] The problem was twofold: MAF's internal disagreements about the best way to proceed and 'the failure on our part to get together a substantial proposal to put before MAF'.[53]

Work continued, but the agendas between the organisations proved divergent. Perry Spiller asked whether the BioGro trademark should simply be sold to MAF, but mentioned also that 'BPC and Biodynamic Farmers & Gardeners have been invited to set minimum NZ standards in conjunction with MAF.'[54] By September 1990 there were '[p]ersistent rumours that MAF are drawing up their own [standards]'.[55] These rumours were aired at a meeting of the NZBPC council with MAFQual staff in November, at which it was confirmed that 'MAFQual is not developing any NZ-wide standards' and instead 'would like to provide a service to assist BPC to deliver the BPC's standards.' The discussion was inconclusive, except that Bob, Chris May and Perry Spiller would continue to negotiate with MAF.[56] MAF's return proposal – that it would do BioGro's work for increased inspection fees – was rejected in early 1991.[57]

This intense focus on MAF really centred on the fact that BioGro was in demand but not financially sustainable and needed government support. This was a direct consequence of the neo-liberal agenda in which government services were expected to turn a profit. It is not at all clear that MAF wanted to dominate the developing organic sector, but it is certainly clear that the organic sector did not want to be dominated and, on balance, seemed to have a preference for managing its own affairs rather than handing over the legacy of the organic movement to a government operating in a paradigm that many felt was incompatible with organic philosophy. After all, the organic movement in New Zealand had developed almost exclusively as a community-driven enterprise. It now wanted the government to embrace its aims and fund the work properly, but on its own terms.

Frustrations with fruitless discussions were keenly felt within the sector. Writing shortly after these events, Perry Spiller remarked,

> There are those within the organic fraternity who feel that if a large hole opened up in Wellington and everything to do with the Ministry of Agriculture from the Minister down to the front doorstep of Gillingham House was to disappear into it, the world and New Zealand would be one hell of a lot better off. Given the treatment meted out before, the same people feel that one breath expended in conversational negotiation with the Ministry of Agriculture and Fisheries is one breath wasted.[58]

Amidst all of this, the government's Health Foods Standards Committee was 'considering the words "organic" and "bio-dynamic" under the Food Act Regulations which if covered would define them for trading purposes'.[59] Some 30 years later this issue would become surprisingly central to discussions on the future of the organic movement, under the leadership of Brendan Hoare.

Although only recently elected to the World Board of IFOAM, Bob was confident that he could bring the biennial IFOAM conference to New Zealand. In January 1990 he wrote to the Christchurch mayor, Vicki Buck, to make his case. He would be attending the IFOAM conference and general assembly in Budapest that August, and while there would like to propose that the 1992 conference be held at Lincoln. 'I put these thoughts to you because I feel you might share the optimism and enthusiasm that this is exactly the opportunity Christchurch has been waiting for.'[60] In April he wrote to Diana Shand, chair of the Canterbury Regional Council's Regional Initiatives Committee, with the same idea; he wrote again in July to tell her that Australian bids to host the conference were being withdrawn in favour of his.[61] In the meantime, he had secured the support of Lincoln University and the DSIR.[62]

His hopes were dashed in Budapest, however: Brazil was chosen for the 1992 event in order to capitalise on the huge opportunity for IFOAM to connect with the United Nations Earth Summit to be held in Rio de Janeiro that year. He complained with the 'strongest possible protest' to the World Board that due process had not been followed in accepting Brazil's late bid.[63] However, the groundwork he had put in lifted his profile and created an opportunity that he was soon able to capitalise on.

Bob was still chair of BioGro's executive committee and the organisation's chief inspector. Developing the inspectorate along professional lines had been a cornerstone of his work for the organisation, but in mid-1991 the demands of this work became too great. In May 1991 he informed the committee that while he was prepared to organise an inspectors' course at Lincoln and to attend the inspectors' meetings, he no longer wanted to carry out inspections himself. The committee asked him to continue as chief inspector, to which he agreed, but he later restated his intention to resign from the role in May 1992.[64]

The following day during the NZBPC Annual General Meeting, Bob resigned from all other offices in order to assess progress from outside the organisation.[65] A number of reasons drove his decision, not least of which was the fact that he had been on the committee since its inception eight years earlier and had spent three of those as chair. Other matters were biting, such as the still-unresolved discussion with MAF about its role; the failure to introduce a producers' levy or to reach a compromise with MAF about a way forward were retarding the organisation's ability to function. MAF's refusal now to offer any of its staff as inspectors unless at commercial rates had had a significant impact on South Island inspections, and Bob predicted that Lincoln's willingness to absorb administrative costs, to say nothing of the Horticulture Department's secretarial time and of course his own work time, would cease.

To add to the pressure, by 1992 funding from the UK branch of the C. Alma Baker Trust had come to an end, as Bob reported to Bernward Geier that September.[66] This placed new constraints on Bob's ability to attend IFOAM meetings and an added drain on his reserves. He was already attending only one meeting in five because of financial constraints.[67] NZBPC needed to make some difficult choices, and perhaps Bob considered that his own and Lincoln's subsidising of the organisation acted as an impediment to progress. He repeated his comment from 1990: 'Nobody should have any doubt in their minds that very soon the Biological Producers Council and its BioGro trademark must become a financially sustainable body. The days of goodwill and volunteers are over.'[68]

The loss of goodwill was certainly true of him, at least. Complaints from BioGro members were increasingly met with frustration. He wrote to one grower in 1991:

> We receive many complaints and criticisms, very few rewards for the voluntary work so many of us do for so many hours a week, year after year … I have no interest in persuading you to remain in the Biological Producers Council, your production is irrelevant and your inspection at present subsidised by our voluntary work. When the true costs of the professional service by MAF are announced you will be one of the first to withdraw your application.[69]

To another member he admitted that he was 'not able or willing to spend the time required to carry out certification procedures to the standard required by IFOAM if we are to take part in the international market for organic product'.[70]

Another factor probably influenced his exit from BioGro. Bill and Madge Crowder recalled their 1982–83 trip to New Zealand fondly and visited again in 1985. They spoke often of emigrating to New Zealand.[71] Bob bought them a return ticket so they could have a good look around and make up their minds – on the proviso that they didn't talk about it again. This trip finally swung it for them: they sold up in Devizes, purchased a house close to Bob in Bowenvale and emigrated in 1990. Their health was failing, however, and caring for them took more of Bob's time.

He was also looking forward to upgrading his 'neglected subversive educational programmes in Biological Husbandry at Lincoln University and the intensification of practical involvement within the Biological Husbandry Unit'.[72] There had been plenty of changes in the BHU. Brendan Hoare left in 1989 and was replaced by Geoff Barnett, who would remain as manager until 1998. The expansion of the unit to 10 hectares in May 1988 meant it was busier than ever. In April 1989 after extensive fallowing, two hectares of the new land was put into mixed herb ley; another two hectares were laid down in a mixed green manure crop. A major shelter belt on the northern boundary of the new area was established in 1988 (the Rose Donaghy shelter belt), and in September 0.3 hectares of onions were sown as part of a project to determine the efficacy of flame weeding. Work on weed control in carrots (by MAF's Rick Zydenbos) proceeded well, as did work (by MAF's Mike Daly) on weed control in wheat in a rotation of beans and amaranthus. Flame weeding trials, which began after Bernward brought flame nozzles to Bob in 1988, had also progressed and became an important research thread, led by Stan le Rooj and later by Charles Merfield. As Bob said, responsibility for maintaining the unit in its dynamic and expanded state now fell on the shoulders of Geoff Barnett, along with Sue Beauchamp in a voluntary capacity.[73] Perhaps thinking of his quarrels with MAF in other areas, he noted, 'All this work has progressed well over the past season and is an excellent example of the type of co-operation that can occur between different organisations when a common philosophy is clearly identified.'[74]

Throughout these years of intense activity in the organic movement, Bob continued to teach. The year 1988, in which Bob turned 50, also marked the final intake of the 'sandwich degree'; the programme Bob had been responsible for would wrap up in 1992. It would be the end to a phase of life that had started for him at Bath University a decade and a half earlier.

Perhaps this explains why Bob was becoming ever more blunt in his critiques of Lincoln – claiming all he received for his efforts was 'ridicule and animosity' – and overtly radical in his teaching aims.[75] In 1992 he proclaimed:

> Teaching is a marvellous way to sow the seeds of radicalism in the younger generation … Subversion is a good thing. I now understand why and how communism infiltrated organisations. Having your people on the inside is very helpful. A lot of the positive governmental reaction to organics is a result of infiltration, and having real believers in there to constantly remind politicians of the arguments.[76]

He had a game plan, and it was out there for the world to see.

One fruit of Bob's approach from this final cohort was Tim Morgan. Tim had grown up with vegetable gardening and studied horticulture at Nelson. Now 20 years old, he found in Bob a 'passionate pioneer, carving out a way in a mire of mediocrity, in economic growth and profit'. He was 'a beacon of hope in a quagmire, an inspiration ... He was a believer ... with a moral compass.' Tim (who changed his name to Sol in 1993) viewed Bob as 'a mentor, a father figure ... He changed my life'.[77] Like Brendan Hoare, Sol saw that 'Bob was not a popular guy [among other academic staff] ... He struggled with the system.' The 'other academics would make the odd scathing comment about Bob ... and I didn't know at the time that there was a lot more to it than that ... I didn't realise his gayness at the time ... and what else was going on for him which was quite large, quite huge ... I know he struggled with it'.[78]

Structural problems within BioGro, and issues at home and in his teaching and the BHU were now compounded by Bob's need to focus on an unprecedented opportunity for organics in New Zealand, which provided the final reason for his withdrawal from BioGro in 1991. At a 1991 World Board meeting in France, IFOAM agreed to bring the international biennial conference to Lincoln in 1994. According to Bob it was 'a great achievement and of immense benefit to New Zealand's potential trading strength as a clean, green country'.[79] He lost no time in drumming up support from players across the primary industries and from the government. In particular he tried to raise funds to allow him to participate in World Board meetings – an ongoing bone of contention but now, with IFOAM '94 looming, more important than ever. He estimated attendance at each meeting cost around $5000. His requests were sent to the New Zealand Kiwifruit Marketing Board, the New Zealand Dairy Board, FORTEX Group, New Zealand Apple & Pear Marketing Board, New Zealand Fruitgrowers Federation, Goodman Fielder Wattie Group, New Zealand Meat Producers' Board, New Zealand Meat Industry Association, New Zealand Trade Development Board, and the Ministers for Environment, Agriculture, Finance and even Social Welfare. New Zealand Apple & Pear Marketing Board responded with a definitive offer; the others declined to support him.[80]

However, Bob's spirits must have lifted when Professor Richard Rowe, who had replaced Mac Morrison as head of horticulture at Lincoln in 1979, wrote to the Fruitgrowers Federation in support of Bob's funding request. Richard said he had 'encouraged Bob Crowder in his involvement with this [organic] movement and his involvement at an international level':

I believe that it is in New Zealand trading interests with food products that this country be seen to be concerned about the health quality of its products ... If some of your members believe that Mr Crowder is an extremist I can assure you they have little understanding of current world attitudes towards chemically free foods and the means of achieving them. Mr Crowder has been a major advocate of New Zealand's ability to produce food products of the highest quality and promoted the requirement for sound scientific research as the means of maintaining that image.

... Mr Crowder in my opinion has done more individually to protect New Zealand's clean green image in world forums than any New Zealander. Any preconceived notion that his activities are designed to undermine our horticulture industries is ill informed and could not be further from the truth. This fact speaks for itself in [the] decision of IFOAM to hold its 1994 meeting in New Zealand.[81]

Rowe's choice of words here confirms the negative view of Bob that had formed in some quarters of the primary sector. Bob was quite capable of standing up for himself, however. Perhaps the best example of this was his war of words with Max Lilley, president of the Vegetable and Potato Growers Federation, which published the widely read *Commercial Grower*. In March 1992 Bob took his class on a field trip to Max's property. It was, Bob told him, 'a disaster'.[82] The negative comments Max made about organics in front of the students had stung; Max claimed they were made 'as a reaction to comments that I had heard from a number of students who felt that they had visited too few properties using conventional growing means and that too much of the course content revolved around organics'.[83] Not so, retorted Bob: the students *only* visited conventional farms.[84]

But Bob's real gripe was that, on top of the 'other bruises and put downs of the last year', Max had written a column in the *Commercial Grower* in which he noted that British organics expert Nic Lampkin, who had been brought to Lincoln to work with MAF, had decided to return home. 'MAF,' Max wrote, 'in its infinite wisdom, has messed it up. Incredibly, Lampkin has no funds to continue his work and he will be returning to England soon, where, less than a year ago, MAF hired him in the first place.'[85] The story of Nic Lampkin's time in New Zealand must be told elsewhere, but Bob was livid. After all, it was on his prompting that the author of the organics bible *Organic Farming* had decided to move to New Zealand. 'Sufficient to say that the loss of Nic Lampkin can be laid not so much at the feet of MAF or the politicians but rather organisations such as your own who have consistently poured scorn on such developments rather than urging support. Had your organisation lobbied as strongly in support of Certified Organic as it has denigrated it, such as during the visit of my B.Hort.Sc. and Comm. class, then we could well be in a better situation today.'

'I do not accept in any way that Vegefed was responsible for the loss of Nic Lampkin,' Max responded.

Bob hit back: 'Vegefed undoubtedly played its part in the loss of Nic Lampkin.'[86]

All of this back and forth occurred within the context of Bob asking the federation for funds to get him to São Paulo for the 1992 IFOAM conference. 'It is not too late however to make amends,' said Bob.[87] The Vegetable and Potato Growers Federation declined the opportunity.[88]

Despite Richard Rowe's support, and although the BHU now had a fully funded manager, there were still problems in resourcing the unit. The BHU was certified organic by BioGro in September 1991, an important step forward.[89] Further, a wide array of research projects was underway there by 1992, enumerated by Bob to Richard in a bid to get research funding: 'I hope the Research Committee will recognise the unit as the valuable asset it is at a time of increasing environmental concern. Nowhere else is there such a unique area available for research projects to be based.'[90]

•

One thing that hadn't changed for Bob was his interest in the weather, which he passed on to his students, as Brendan Hoare, Jared White and Tim Jenkins all recalled vividly. It was particularly interesting to observe how well BHU soils now absorbed heavy rain, as compared to conventional neighbours. More dramatically, a pall of dust from the eruptions of Mt Pinatubo in June 1991 and Mt Hudson in August 1991 covered the skies and sent temperatures plummeting in autumn 1992, causing crop damage and early frosts. In August that year 'the Big Snow' shut Christchurch for days and farmers suffered tremendous losses with spring lambing. Bob set out the pattern of heat units throughout much of the twentieth century in an article in *Growing Today*. He commented, perhaps wryly, that it was 'not surprising the Canterbury population noticed the difference and felt CO_2 warming had come to an end. They're jumping the gun of course: long-term departures from expected mean values are not foreign phenomena to Canterbury.'[91] The Big Snow was also noteworthy because it forced the postponement of the inaugural meeting of the Central Organising Committee for the tenth international IFOAM conference, which by now was consuming much of Bob's life.[92]

Rows of healthy vegetables, the result of successful crop rotation.

7: IFOAM '94

THE NEW ZEALAND GOVERNMENT'S POSITION ON ORGANICS must be seen in the context of a wider international policy discussion. The government had embarked on a sustainable development agenda, which commenced under the previous Labour government with the creation of a Ministry for the Environment in 1986. By 1991 this had led to an overhaul of development legislation in the form of the Resource Management Act. The movement towards an appreciation of 'sustainable development' was boosted by the landmark 1992 United Nations Earth Summit in Rio de Janeiro, from which flowed so many international environmental agreements.

This helps explain MAF's frustration with BioGro, which it had tried to skirt around with Bernward Geier in 1989. Led (one could argue) by socialist philosophers, BioGro was not well positioned to rise to an opportunity presenting in a neo-liberal context. It refused to charge commercial rates for its services – the market was still too immature – and as a result was not financially sustainable. Bob's own pragmatism and willingness to compromise with farmers on specific certification issues was clearly not enough to shift this dial. He lamented to the minister of agriculture in August 1991 that relations between MAF and BioGro 'appear to have deteriorated'. This, 'while the potential for Certified Organic world trade goes on rising rapidly in all our major overseas markets'.[1] Nevertheless, due in part to Bob's manoeuvring at IFOAM (which directed the government back to Bob), there was a stalemate.

MAF's policy paper of May 1991 said that the ministry should, among other things, 'assist the development of national standards for organic production'.[2] But by September that view had been reversed: 'MAF believes that the current system of private standards is performing adequately and will therefore not move to establish uniform national standards for organic farming and organically grown food.'[3] It did not rule this out for the future, however, but it did propose a number of policy actions that must have been welcomed by the sector. These included developing a formal policy of encouragement and support for organic agriculture; supporting public-good funding for organic farming research and technology transfer for 1992, 1993 and beyond; supporting acceptance by foreign governments of uniform international organics standards; and facilitating exports of organically grown produce by certifying products as meeting foreign standards.[4] Not too bad for a sector that in 1991 was valued at only $1.5m – less than 0.1% of New Zealand's total agricultural production.[5]

MAF's final policy position paper on organic farming, 'Towards sustainable agriculture, organic farming' (1993), mentioned IFOAM: 'Bob Crowder from Lincoln University has been the New Zealand representative and has sought to broaden IFOAM's outlook beyond Europe.' It noted that the upcoming conference in Lincoln:

> provides an excellent opportunity for New Zealand to obtain international recognition as a producer of organic products. It will also be an opportunity for the promotion of more sustainable agriculture, and specifically organic agriculture, as a viable alternative farming method. The Minister of Agriculture has already expressed his support for this conference and has encouraged various departments and businesses to provide financial backing.[6]

The paper confirmed that the government believed organic certification was best handled by the private sector and, once again, would not be moving to establish a national standard.

In early 1993 Bob travelled to the UK to attend the British Organic Conference at the Royal Agricultural College in Cirencester. He was intrigued by the 'major consideration' given to the interaction of the European Economic Community policies with conservation and organics groups and noted in his diary, 'the EEC details indicate a real concern for the issues'.[7] He was disappointed with his own presentation, which he felt was unconvincing, even though the response was positive.[8] Perhaps he felt pressure to excel, given the upcoming Lincoln conference. He recorded, 'Good discussions with many people over the conference in 1994.'[9]

Following the conference he set out to travel with Martin Stokes (an old flame he had met in Bath 1975). He was surprised in Devizes when their host, Mrs Megraw, 'hoped Martin & I would not mind sleeping in separate rooms!! O well, perhaps people know more than we give them credit for.'[10] Together they visited Doubleday and Derby before parting at Stanstead Airport, where Bob boarded a plane for Frankfurt. He wrote, 'Martin has been such a good companion over the last few days.' It must have been hard to say goodbye.[11]

From Frankfurt he travelled south to 'the enchanted village' of Heppenheim: 'Quiet, cold, tranquil, smoke rising lazily over houses. What a contrast to new areas but went up to the castle on top of hill past vineyards, woodpeckers, hawks circling – world pulsating just close by.'[12] At the Bio-Fach International Organic Trade Fair in Wiesbaden he was reunited with Bernward Geier and others from the IFOAM board and busied himself setting up his display and exploring. 'Words cannot describe this exhibition which encompasses everything from food to textiles.' He was bemused by the 'well-dressed, healthy people smoking incessantly at the entrance where a great round "ash tray" existed' and made contacts that could prove useful in New Zealand: customers wanting wool, amaranthus and honey.[13]

From Bio-Fach in Wiesbaden, events shifted to Theley and an important awards night for IFOAM: Bob called it the 'big evening … rather ceremonial but pretty significant'. The ministers of the environment for Saarland and the German government were present, making this a 'really benchmark event'. Bob was surprised to receive a certificate of recognition from IFOAM, along with Hardy Vogtmann and Pierre Ott – both key players in the development of the international organic system.[14]

The New Zealand press picked up the media release about this award. Bob was acknowledged as 'a pioneering figure in the New Zealand organic husbandry movement'. The media reported on his trip, spoke of Bob's optimism for organics, and noted that Lincoln University would host the IFOAM conference in 1994.[15]

Bob's spirits were lifted by the excitement and sense of progress – and purpose – that emerged from the evening at Theley. The next day he attended the IFOAM World Board meeting: the board worked well as a team and he felt a sense of respect for the other members. It was, he wrote, 'very important to be present today' – not least because somehow a 'strong feeling' had developed that New Zealand 'was being arrogant over the conference [to be held there the following year] … I hope I have dispelled this feeling'.[16]

He travelled next to an organic conference at Asilomar State Marine Reserve, just around Monterey Bay from Santa Cruz. The crowds were 'very organic – NO smoking evident'. A rap artist with a 'radical message' opened the event to 'standing rapture applause', and Ralph Paige offered 'an excellent message for small-scale production by [the] underprivileged'. He was inspired by Ben Cohen of Ben & Jerry's ice cream, whose ethical message marked him in Bob's mind as a possible feature speaker for IFOAM '94. Miguel Altieri's message about indigenous sustainability also met with his approval.[17]

But while the small-farming and indigenous community message came through in the opening – quite the turnaround from the Californian system Bob had encountered in 1969 – the conference also featured discussion on 'bio-tech', also called 'recombinant genetics' and later 'genetic engineering'. Bob took notes on new trials coming out of UC Berkeley in which the 'cutting and splicing of genes' resulted in squash plants that were protected against four kinds of virus, a *Bacillus thuringiensis* (BT) gene protected them from insects, and another gene could block ripening in tomatoes so that the 'tomato will remain green on plant for 150 days or until *Ethylene* is applied'. According to the speaker, modifications of potato genes to increase starch were justified because the potatoes required 'less oil for chipping'.

The 'other side of the story', which he said 'was good to hear', was then put by Katherine Griffith. The US government would spend $10.5m to help increase herbicidal tolerance by 25 percent. Fifteen percent of applications for approval were essentially BT modifications that were causing resistance to normal BT applications; outbreeding of resistance into weeds could occur; there were health considerations from the toxin actually being part of the plant; and the high crop losses being remedied through these bio-tech solutions were, Bob thought, covering up 'implications of *POOR husbandry* practice'. There were moral questions about patents and royalties for new genetic combinations, including the offspring from animals, and, last but not least, consumer faith. Eighty-five percent of consumers wanted labelling on genetically modified foods, but industry was resisting: why then should public money be funding this work? Bob found Griffith's critique 'very good' and the workshop 'interesting and excellently balanced'.[18] That evening he dined with Allan Savory, whose theories around brittle and non-brittle environments stimulated Bob's thinking.

Bob was gathering ideas to make the Lincoln conference memorable: 'An excellent conference covered good science and practice right down to spiritual concepts. An excellent blend of people covering diverse interests from the best science to the folksyest people. It all sat well together and made for an excellent experience.'[19] He took home a swag of ideas and contacts to help bring the 1994 conference to life.

•

Bob's relationship with certain members of Lincoln's academic staff continued to deteriorate. Others supported him publicly, though articles such as Roland Clark's 1993 piece in *Growing Today* cannot have pleased the university management. Clark wrote that Bob was selling onions from the BHU to a city supermarket and 'getting three to four times as much as conventional growers'. And although he had been 'ridiculed',

> fortunately Bob Crowder's skin must be as impervious to insults as an elephant's to a flea, because Bob just went on developing his 'weird' ideas in the face of all critics ... Incredibly, unbelievably, Bob has never managed to persuade, bribe or threaten any of the world-class animal scientists at Lincoln to conduct a trial on these [BHU] pastures. Damn it, these so-called seekers after truth are missing

a golden chance of at least proving he is talking a load of rubbish! Or not, and perhaps that's why they try to pretend the unit is not there.[20]

According to Clark, Bob's work at Lincoln was outstanding. The orchard contained 80 varieties of apple that had received no artificial inputs for years; at least six of the old varieties were 'unblemished' – an excellent result. The floriferous understorey of *Apiaceae* (previously called *Umbelliferae*; the carrot family) and a plethora of other flowering plants attracted predators of various pest species. Above this was a shrub layer that included flax, kōwhai, raspberries and gooseberries, and above these were shelter trees such as Lombardy poplars for coppicing. Then there were the vegetables. 'I was literally staggered when he brought me to his organic cropping area,' wrote Clark. 'I have never seen such healthy-looking crops. Never.' He described the six-year rotation of the semi-intensive beds: rest for a year with ryegrass, oats, mustard and clover, then brassicas, potatoes, amaranthus, garlic and leeks, corn, and finally sunflowers with squash as the understorey. 'I have to say it again: these crops looked marvellous and they had never seen a spray.'[21]

Despite the difficulties he encountered in his work, Bob continued to speak and write enthusiastically about the abundance of life at the BHU. In *Growing Today* he wrote, 'Our experience over 16 years has been [that] some of the best things in life are free … an environment carefully created on the basis of sustainability through diversity will allow "care-free" productivity in the true sense of the word.' He particularly promoted the use of flowers in the growing system. 'Incorporating flowering plants into horticultural and agricultural regimes provides habitats for beneficial insects that fit into integrated pest management programmes and help maintain a natural balance without chemical intervention':

> Building a sustainable system based on organic techniques involves the bringing together of many parts to make a whole. The floriferous nature of that whole is a vital part in the system and works best when incorporated with a total holistic appreciation.
>
> At the Biological Husbandry Unit there has always been a conscious effort to have flowers throughout the growing season, especially the umbelliferae, and the greatest effort has gone into creating the now well-known 'Floriferous, Umbelliferous' understorey for orchards.[22]

His system was working and was simply *better*. Better for soils, for biodiversity, even perhaps for the spirit as it created 'a more aesthetic environment both for work and play … Environmental strength through diversity is the future pathway for sustainable agriculture in New Zealand.'[23]

•

The wheel of life for his parents, now 85, was turning. On 6 June 1993 Bill died from kidney failure and Bob became Madge's primary care-giver.[24] He organised live-in help for her to begin with, but eventually when Madge's needs increased she moved into the retirement home next to his house on Ashgrove Terrace. Bob built a gate in the fence so she could visit his garden easily.[25]

The BHU continued to be supported by the enthusiasm of Bob's students and many volunteers. While Sol Morgan was doing his honours project on the BHU Granny Smith orchard in 1993, he and another student, Jared White, looked after the Wwoofers together. Jared had arrived as a 19-year-old in 1990 specifically to Wwoof at the BHU, and Bob convinced him to stay on to do his degree, which he completed in 1993; he remained at Lincoln as a tutor until 1995. Jared was 'immediately drawn to the place and the people, and there were so many great things to learn':

> I was immediately drawn to [Bob]. He is very charismatic … I was … under his spell with his enthusiasm for organics … I remember the ways he would teach the Wwoofers and us how to do things, and he was really good at not assuming that people knew anything. So he would explain to people how to use hand tools … and how to plant a seedling … It was really good learning … he took me under his wing.[26]

Looking after the Wwoofers meant being a stable presence in the Wwoof house down the road from the BHU. Sol and Jared both lived there, and there were two spare rooms available. They made sure the house ran smoothly and that there was enough food.

Wwoofers were an important fixture of the BHU that circled back, in a way, to the nature of horticultural labour issues that had interested Bob in the 1970s. Wwoofers are not paid to work but receive an enriching experience in return for their labour. At the BHU they were expected to stay for around two weeks and worked four full days in seven rather than the four hours per day suggested in Wwoof guidelines.[27] Bob screened them carefully and only accepted those who were philosophically committed and motivated environmentalists.

Stuart Jeffrey and Page Lawson fitted this description when they arrived at Lincoln from Scotland for the IFOAM conference in 1994, hoping to get a Wwoof spot at the BHU – the 'Mecca of organics' as they saw it – which they had read about in *Soil & Health*.[28] Having plucked up courage to apply to Bob, they started Wwoofing at the BHU immediately after the conference. The Wwoof accommodation was full – even an office in the BHU building set up as a bedroom was in use – so Stuart and Page camped in the Rose Donaghy shelter belt. 'It was easily, far and away the best place I ever Wwoofed,' Stuart said. 'If there was a job to be done, we all got on and did it … Five acres of onions? We'll all weed the onions … It felt like what I thought organics should be. Extremely generous, and kind, and equitable.'

Jared working in the corn, BHU, 1990–91. (Photo courtesy of Jared White).

> The one thing that's been with me my whole life since then, is that, oh my God, this is possible … Walking down into … the intensive garden … and the apple orchard … it was like what I'd read about in a book, but it was real! These towering sweetcorn and beans and tomatoes and peppers, and we could just walk up and down and pick anything we wanted.
>
> … The BHU seemed to be a place where people weren't necessarily afraid to give something a try … And in conversations with Bob … about his visions for the place, what he pictured that place being able to become. An eco-village, basically. Not just a place where trials were done, but a place where people lived and worked, and where wind was harvested, and earth buildings were built, and aquaculture was practised …
>
> It felt to me like a kind of Utopian democracy because everybody did everything together, and equitably. That's what it … exemplified as much as I'd ever encountered, and as much as I've ever encountered anywhere … because it wasn't a commercial place.[29]

A small team of volunteers from the wider organic movement also helped keep the BHU going. They came whenever they could, meeting in the field at the BHU and often spending their time on the intensive beds – that nucleus of the BHU that Tim Maples had been paid to establish so many years earlier. The lively Sue Beauchamp had started volunteering in the 1980s. Nicole Bührs was another, an analyst programmer and organic gardener who had been a member of Soil & Health since 1985 and came to Lincoln from Auckland in 1991. She paid a visit to the BHU, met Geoff and Bob and was impressed. After that she volunteered one day per week for the next seven years, reasoning that she would learn how to garden organically in the radically different climate. 'I met the whole organic community there, because Bob saw it as a demonstration area, and people came … I made real friends.'[30] These friendships built and strengthened an expanding network of change-makers; in Nicole's case it led to the development of the Canterbury Permaculture Group and work in adult education with the Organic Garden City Trust.

Charles 'Merf' Merfield also came to the BHU at this time. Having worked at Weleda in Hawke's Bay, he arrived in Canterbury in 1994 and Wwoofed with Tim Chamberlain and David Musgrave at their properties. Charles was considering further education at Lincoln and was drawn to the BHU after seeing it listed in the Wwoof directory. He began a postgraduate diploma and became involved in the BHU through the student group, Lincoln Environment Organisation.[31] His involvement continued in various capacities for over 25 years.

•

Italian Wwoofers with Helen Duckworth (centre) and Nicole Bührs (right).

The tenth international IFOAM conference, with the theme 'People, Ecology, Agriculture', was planned for 11–14 December 1994, followed by the IFOAM General Assembly on 15–16 December.[32] IFOAM operated in ways not dissimilar to organisations such as Soil & Health and the NZBPC: everybody was stretched beyond their means and the whole thing was kept afloat by sheer determination and loving mutual support. Bernward was finding the logistics of herding IFOAM's many cats taxing, but Bob reassured him: 'everybody loves you and I am sure we will all sit up and take notice'.[33] Bernward faxed Bob in February 1994 to congratulate him and conference organiser Don Crabb on their tremendous efforts to date: 'If you continue like that, you are seriously challenging for the best ever IFOAM conference. I do not just say it – I mean it. But this is not a real surprise to me, having already experienced some years ago [in 1988] the excellence and capacity your movement has to organise conferences.'[34]

Bob continued to drum up support on the international stage. During an IFOAM World Board meeting in February 1994 in Germany, he sensed '[g]ood vibrations concerning the conference – no problems foreseen. Must keep people up to the promises they have made.'[35] One issue the IFOAM '94 team debated at length was whether to produce proceedings from the conference. Bob consulted both Bernward and Nic Lampkin on this matter and in the end decided against this enormous undertaking. On his return to New Zealand Bob wrote to Bernward, imploring him to '[p]lease have a good rest yourself and *Thank you* for your hospitality under such stress'.[36]

Correspondence from this time shows Bob was very much at the centre of the organising, handling endless requests and being eminently diplomatic in his responses.[37] The whole New Zealand organic movement got behind the conference, and a large team assisted Bob with planning and coordination throughout 1992, 1993 and 1994.[38] It was an enormous undertaking. Pre-conference tours of 'historic, scenic, conservation and tourist attractions as well as places of organic significance' were organised for Auckland, Bay of Plenty, Hawke's Bay and Nelson, as well as in Australia from Cairns to Adelaide.[39] There was to be a public 'Fayre' and grand opening at the town hall, and the conference itself included speakers, posters and workshops. By January 1994, 160 papers had been submitted; in September this had risen to over 200.[40]

Prince Charles (now King Charles III) was expected to visit New Zealand in February 1994. Invited to the reception for the Prince at Government House, Bob seized the opportunity to present him with an invitation to open the conference. The Prince demurred but promised to send a suitable message to be read at the opening. His message acknowledged the timeliness of the conference:

> ... as many countries throughout the world are thinking about implementing sustainable development agreements reached at the Rio Conference [in 1992] ... my own view is simply that organic farming is as close as we are going to get to sustainable agriculture. It is, after all, remarkably close to the traditional farming systems which have proved themselves to be sustainable over hundreds of years, until the advent of the high input–high output, technologically driven, agri-business approach to farming.

The Prince praised the work of IFOAM, whose 500 member organisations in 95 countries were 'a remarkable achievement and an example of international co-operation in agriculture almost without parallel': 'Unless we can develop systems of food production which work in harmony with the natural environment, the future for our planet and, therefore, for our own descendants must surely be bleak. Organic farming is a system which has precisely these attributes.'[41]

Bob anticipated the conference would not be the 'usual, everyday type of dry, smug insiders conference ... IFOAM '94 will ... have its philosophers and practitioners. People of the land, with dirt under their fingernails, all coming together with one common understanding: that there must be another pathway for the world to tread.'[42] Hosting the conference was 'a real coup for New Zealand. One that offers us great opportunities to put ourselves on the map as a true clean, green producer'. The conference was 'a reflection of the increasingly mainstream nature of organic and sustainable agriculture in this country':

IFOAM is about people, how they live and treat their environment, and about their agriculture and how it interacts with society. Philosophy is about thought, wisdom and understanding and IFOAM was established by people who have tried to take a broad view of life. That philosophy, and how it can be translated into practical measures, will be the focus of IFOAM '94.[43]

The event opened at midday with an Organic Fayre in Victoria Square, 'an attempt at "outreach" to the general public to make them aware of the environmental issues to be discussed … and that these issues are not just for academics but involve them'. There was a range of food and information stalls and entertainment and, of course, a performance by the Lincoln Tussock Jumpers (as Bob's Morris side was now called).[44]

The official opening was held at Christchurch Town Hall and welcomed the public as well as delegates.[45] After a pōwhiri, conference patron and chair of proceedings Sir Peter Elworthy got things underway.[46] In her welcome address, Mayor Vicki Buck declared her hope that Christchurch could become the world's first organic Garden City – a concept that would gather momentum in the coming years.[47] She was followed by IFOAM president Tom Harding, Associate Minister of Agriculture Denis Marshall, Bernward Geier, and Helen Browning, who read the message from the Prince of Wales. Peter Ellyard's address, 'Clean green future: Creating a sustainable society', was followed by physicist and ecologist Vandana Shiva's speech on 'Free trade versus freedom' and 'the implications of GATT [the General Agreement on Tariffs and Trades] for world sustainability'.[48]

Helen Browning's presentation demonstrated that IFOAM was engaged in much more than a different system of growing food and fibre. Reflecting on the British experience, she pointed out that in both urban and rural areas 'there is a realisation that our greatest loss over the previous forty years, has been the erosion of our communities':

> The development of 'new economic' thinking, which I am sure we will hear a great deal more about here over the next few days, will have a profound effect on the way we view the world and its resources. This, I believe, is our only real weapon against the potential havoc that GATT … threatens.
>
> … The organic movement must be part of this wider shift towards a saner world … [It] is very encouraging that this conference has as its main theme the connections between people, ecology and agriculture. As organic producers, we can demonstrate theory in practice and be the engine (literally) behind the forces for change. Agriculture is fundamental to our civilisation, and without sound foundations any new era will be built on sand … There is a fundamental need for us to widen our perspective, to reach out to all levels of society with our practical and far-reaching philosophy. I believe that this conference is about exactly that.[49]

Promoting IFOAM '94 in Victoria Square: Bob and fellow Morris dancers performed in their Tussock Jumpers outfits, and organic produce from the BHU was on display.

Bob's hard work and focus in networking from 1991 through to 1994 had resulted in a stellar line-up of speakers, a real who's who of the organic and environmental world, including not only Vandana Shiva, but also Hardy Vogtmann, Allan Savory, Miguel Alteiri, Stuart Hill, Lawrence Woodward, Nic Lampkin and many other top-echelon organics experts and environmentalists.

Bob had curated an event that spoke to all his own values; even his spiritual side was given space in Erihapeti Rehu-Murchie's address on sustainability, environmentalism and spirituality from a Māori perspective, and Professor Lloyd Geering's 'linking environmental values as the basis of a new spirituality and ethic in the world of the future'.[50] Roland Clark, writing about the conference for *Growing Today*, said, 'Geering's idea that we need a new ecological Christianity open to all horizons of human life, open to other religions and infused with a desire to affirm our inseparability from the natural world satisfies me deeply.'[51] It satisfied Bob as well – so much more than his early experiences of church life.

IFOAM '94 was a tremendous success. New Zealand was in the spotlight of the international organics community, and it shone. Bob summarised the event in 1995: 'It has been acknowledged as the biggest international conference at Lincoln University to date, a great and influential success both at home and overseas and handled over 800 persons and over $500,000.'[52]

The conference brought together a number of people who would play important roles with the BHU, including Lady Fiona Elworthy and her husband, Sir Peter; Jon Manhire; Morgan Williams (then with MAF Policy but soon to become the second parliamentary commissioner for the environment); Mark Hazeldine (NZ Trade Development Board); Peter Townsend (Canterbury Development Corporation); and Allan Gillingham (AgResearch).[53]

A report from the BHU emphasised that the 'IFOAM Conference secretary, Don Crabb, is of the opinion that without Bob Crowder's influence and presence on campus, the IFOAM Conference of late 1994 would not have been hosted by Lincoln University.' The Student Services Division of the university grossed $250,000 in income from provision of meals and accommodation to delegates, while the university received approximately $24,000 worth of publicity from print media alone. 'Although television coverage was not as extensive as that received in print, the value to the university is almost certainly higher.'[54]

The 1994 IFOAM Conference at Lincoln made an enduring impact on the organic community in New Zealand. To Bob, it was the 'pinnacle' of his career.[55]

Bob, Murray and Jill in their Ashgrove Terrace driveway, 1989.

8: The struggle to save the BHU, 1995–98

BOB BEGAN 1995 with a visit to Perth to see his friends Rod and Ted whom he'd met on the *Southern Cross* in 1962 and while there judged a drag queen contest at the Court Hotel. His friends' life was 'dramatic and plush'. Bob enjoyed the vitality of the nightclubs and remarked that he 'did not feel tired, sick, or stressed. New lease of life, in fact justifying making the effort to get out and about. Feeling quite pleased and confident for a change.'[1] These comments, and his uncharacteristic enthusiasm for the gay scene, reflected his relief at escaping from the challenges of work and home life.

The visit to Perth was part of a longer trip to Britain that was very much about his parents. Since her husband's death in 1993, Madge now lived in a retirement village next door to Bob and suffered from arthritis and other ailments.[2] She missed Bill and wanted to return to England to be near her friends. Bob spent some time exploring nursing homes and retirement complexes in Devizes but was not impressed by the options: 'A step back to Oliver Twist', 'sparse furnishings' – even the 'best of bunch' was still a 'pale shadow of Cashmere View', where Madge was living.[3]

While in Devizes, Bob scattered his father's ashes on Roundway Hill. The day started with heavy rain, which cleared in time for the ceremony; on Roundway it was dry with a fresh westerly. Afterwards, the small group of family and friends ventured to the Bear Hotel for coffee and cakes and on to Bowden Hill to see the snowdrops and primroses.

Bob stayed with his former partner Chris Weeks and Alan, Chris's current partner.[4] In Christchurch, his relationship with Murray – his partner since meeting at the Green House back in 1982 – had not been going well. Bob rang him from Bonn but found the call 'a bit depressing all around.'[5] Murray would leave Bob later in the year, bringing the roughly 12-year relationship to a sudden close.[6]

Bob was becoming disenchanted with IFOAM. The event he was in Bonn for was 'boring, boring and entirely badly managed. Need to assess what is wanted in a conference.' He found some inspiration in the Bio-Fach Trade Fair, but an IFOAM strategy seminar in Theley was 'depressing, meaningless – waste of time, endless'. Although the 1996 conference, held in Copenhagen, provided some inspiration, Bob's notes nevertheless concluded with the question that many conference-goers ask: 'BUT, but, BUT?'[7] Where was all this activity leading?

None of this helped with the come-down that had followed the success of the 1994 conference. Although the BHU was recognised nationally and internationally as an important organic demonstration and research station, and those connected with it continued to build knowledge and confidence under Bob's leadership, Lincoln University still had not embraced it fully and would not confirm plans for the future. Nicole Bührs remarked after IFOAM '94, 'We thought, now, the university will finally see that this place is booming and that people are interested, and they are going to pour money in there. And nothing happened. We went from the top of the mountain down very fast … We were very disappointed.'[8]

Bob was frustrated too:

> I thought after the IFOAM conference the university would see the advantage of having a thing like the BHU and that the government would … take advantage of having had all these leading world scientists in the country … but the media kind of ignored it … it was all a bitter disappointment, really, the aftermath of IFOAM.[9]

The unit struggled financially. Lincoln's contribution allowed for only bare essentials, and Bob gleaned extra income from research grants and produce sales. Sue Beauchamp ran a weekly market on Lincoln campus for years, and produce was sold at the BHU and to various shops and restaurants.[10] Bob supplied vegetables and fruit to the New World supermarket in St Martins, where a nascent organic food section had begun with sales of eggs and chicken from Ernst Frei.[11] Aided by the store's owner, Brian Newbery (who became a champion of the GE Free movement) and later by Rod Donald MP (whose Green Party office was next door), the organic section – the first in a New Zealand supermarket – slowly grew to become a point of difference.

•

Lincoln University established a review group to assess the BHU and make recommendations about its future. Bob was asked to provide an overview of the unit, including information on the input from volunteers in recent years, a publications list, details of how the BHU was used in teaching, and its sources of income.[12] Volunteers were vital, Bob replied: 'without them the unit would not have the standing it does on the world organic scene'. He described Sue Beauchamp's significant voluntary contribution and noted that in the three months before the conference there were up to four volunteers working almost full time. In terms of publications, many articles had appeared in the farming and popular press, and Bob reminded Lincoln that he was not employed to publish peer-reviewed journal articles. He pointed out that the unit was used for research by others, such as Trevor Jackson and Mike Daly – both of MAF. He described how the unit had been part of the horticultural degree programme since 1976 and noted that numerous short courses were conducted from the BHU, including for the Enviroschools programme and adult education programmes. The only external source of income was the Soil & Health Association, which, he estimated, had contributed close to $50,000 dollars over the years.[13]

The university's final report when it came acknowledged the relevance of the BHU to Lincoln but contained caveats. 'The unit provides significant benefits to the university, principally from public relations profiles and publicity but also from showing that the university is active in this strategically important area. The direct costs for staff, land use etc are higher than the direct financial benefits, but recognition must be given to the intangibles which are impossible to cost accurately.'[14] Although the report highlighted the roles of demonstration, teaching and production of specialist crops for sale, the writers considered the main opportunity resided in research. Social issues of biological husbandry might attract research funding, but more important were ecological studies and 'underpinning research for organic production systems', which was supported by industry.[15] The report recommended that the acreage of the BHU be 'adjusted in line with the research demand identified in the [proposed] business plan', and that 'the present academic 1 FTE [Full Time Equivalent staff member] be evaluated in relation to a new contract for appropriate consulting to the BHU, consulting on PR work, teaching and other possible roles and that this be job-sized formally'.[16]

Bob found the contents of the report discouraging and was 'bitter about the whole way things were going'.[17] Allied to this was his position on the IFOAM World Board, from which he drew so much mana. Bob funded his work on the board largely from his own pocket. In 1995, in an attempt to leverage off the success of IFOAM '94, he sought funding from organisations that had been involved in and derived some benefit from the conference but had not sponsored it. However, none of the parties approached came to the table.

With the continued lack of direction from Lincoln management, the organic community again stepped up to support the BHU. This grassroots support base had been given a fillip from the IFOAM conference, and particularly from the proposal by Christchurch Mayor Vicki Buck that the Garden City of Christchurch could become the first organic garden city in the world. Bob wrote to her early in 1995, thanking her for her participation in the conference and adding, 'It was pleasing to hear you talk at the opening, of the resolve of the Council under your leadership to create an "organic" city':

It is interesting to reflect that in many advanced societies a very large proportion of the inputs used in agriculture are in fact used in urban areas on gardens and amenities for recreation and sport. The initiative within the Botanic Gardens to cut down on such inputs is welcome. If I can be of assistance in helping to hasten the 'greening' process city wide, then I shall be happy to cooperate.[18]

Vicki Buck's call led to an 'Organic Christchurch Campaign' in 1996, which brought together Christchurch City Council, the Canterbury branch of Soil & Health, the Sustainable Cities Trust, Rod Donald MP and others, to form what in 1997 became the Organic Garden City Trust (OGCT). This in turn became a legal umbrella for an array of project groups that included Kids' Edible Gardens (delivering organic gardening programmes to primary schools), the Community & Home Gardens Group, which soon morphed into the Christchurch (later Canterbury) Community Gardens Association, and the Canterbury Commercial Organics Group (CCOG, or at first the OGCT Commercial Group). This last had its antecedent in the 1980s as the Canterbury Organic Producers, discussed earlier.[19] Bob was involved in all of this.

This surge of interest in organics reflected a national trend. In 1996 the Green Party gained three seats in Parliament as part of a broader coalition of left-wing progressive parties under the umbrella of the Alliance Party. Green Party MP Rod Donald wrote to Bob in 1997 asking him to 'keep me up with any issues which need to be pushed at the Parliamentary level … Do make the most of me being here.'[20] Jeanette Fitzsimons, now also a Green MP, provided an example of how this support might look: in question time she asked the minister of agriculture if his ministry was undertaking any work towards a national organic standard, or towards ministry support for the existing BioGro and Demeter standards.[21] The ministry was not.

Surprisingly, perhaps, the new National Party–New Zealand First coalition government agreed to set aside $30m towards organics research, a move Rod Donald referred to as 'one of the really innovative commitments contained in the coalition agreement'.[22] Bob was delighted to learn of this and wrote to the Minister of Agriculture, Lockwood Smith, expressing his pleasure. He hoped this would mark 'the end of an era when certified organics has appeared to be marginalised in favour of the much less well defined concept of sustainable agriculture'.[23] Any hopes the organic movement may have had for a breakthrough were dashed, however, when the minister emphasised the wording of the agreement – '*up to* $30m', not necessarily the full amount – and pointed out that much of the research already being undertaken through the Public Good Science Fund was 'equally relevant to organic and conventional primary production'.[24] In short, little was about to change.

Delegates at the IFOAM World Board meeting, Theley, 1996. Bob (fourth from left), Bernward Geier (far left).

'Bob – obviously we will be pursuing this copout!' Rod Donald wrote in red ink on his copy of the minister's letter.[25]

Bob pursued it as well, berating the minister: 'Clearly the reference in the Coalition Agreement is for additional spending on the organic system. It does appear as if the government has once again decided to go back on its promise and has never intended to spend an extra $30 million on organic agriculture. This is most regrettable as New Zealand is already losing some of its clean, green credibility in the world.'[26]

•

Fed up with the lack of direction offered by Lincoln, Bob announced his retirement from lecturing in July 1997. *Commercial Grower* shared the news in a review of his career. The article began, stingingly:

> It would be fair to say that Bob Crowder, Lincoln University's recently retired lecturer in organic production, has never been comfortable with conventional production. And it would be equally fair to say that his often abrasive and dogmatic attitude towards anyone who does not agree with him has not encouraged much sympathetic support in the industry.[27]

The article was not all bad, however. It acknowledged the IFOAM Recognition Award Bob received in 1993 for his 'international contribution to biological husbandry', and the fact that, due to the success of the 1994 IFOAM conference, 'Lincoln's international reputation for organics expertise was confirmed.'[28]

Bob had been working at Lincoln since 1966 and his retirement created a natural moment for reflection on his achievements. A coherent narrative about the BHU's origins began to emerge.

> Yes, it is 21 years ago that Tim Maples, a Horticulture Diploma student at Lincoln University, then College, initiated a radical student call for more environmental, sustainable organic input into courses at the college ... Hard to remember the agony and the ecstasy now. Hard to understand that 'Biological Husbandry Unit' was chosen as a name because of the need to divorce the area from a much discredited and ridiculed word, 'ORGANIC'.[29]

Bob reflected that organics had made progress over the last 21 years through the term 'certified organic' and the development of a commercial organic industry. However, this was only 'in so much that it sustains the concept of large-scale production for export within the existing economic paradigm'. The BHU, which existed outside of this, provided 'hope for those who see that the future must depend upon the ability of society to nurture itself not only by food and drink but also through the spirit'. He warned: 'The BHU is in danger of being marginalised yet again in favour of Trade/Export-oriented conversion of a university farm to Certified Organics using some of the reinvention of the wheel: relearned technology but jettisoning the philosophy.'[30]

It was a difficult time for Bob. For distraction, and validation, there was always the international organic movement, even if sometimes this work involved long afternoons of what he called 'irrelevant talk' interspersed with 'good outcomes'. In Theley, Germany, in March 1997 Bob chaired 'the historic meeting between Accreditation & World Boards which resolved differences and created the new legal identity of Accreditation'. He felt he handled it well and noted in his diary, 'Perry Spiller will be pleased to see the result of a simple IFOAM Accredited label.'[31]

On 2 June 1997 Bob received formal recognition for his part in the organic movement in New Zealand: he was made a Member of the Order of Merit 'for services to organic husbandry'.[32] Tim Chamberlain organised the submission, and the curriculum vitae he prepared for Bob was impressive: it traced his work as a young man within the established scientific community, his growing interest from 1976 in climate risks and climate change, and his 'Think Big, Naturally' message of 1982. Tim commented,

Bob becoming a Member of the New Zealand Order of Merit, 1997. Left to right: Jim Kebble, Bob, Mary Zohrab, Rt Hon Sir Michael Hardie Boys, Rose Donaghy, Tim Chamberlain.

Bob Crowder believes that the creeping malaise of world pollution is at present making a stunning impact in many areas of the world. New Zealand has an unassailable advantage, through its world isolation, to use what previously has been a disadvantage in showing the world how to develop a clean, non-polluting, self-sustaining, healthy, productive and prosperous society.

He believes in the motto Food You Can Trust grown the natural way in the pristine purity of the South Pacific.[33]

Tim quoted Sir Peter Elworthy's 1995 words:

'Our New Zealand success in organic farming can be directly attributed to the efforts of Mr Crowder in his development of the Biological Husbandry Unit at Lincoln, his early work with the Biological Producers and Consumers (now BioGro New Zealand), his work with the International Federation of Organic Agricultural Movements (IFOAM) which concluded in the triumph of the IFOAM International Conference at Lincoln late last year. At that conference, Mr Crowder was publicly identified and recognised as the founder of organic farming in New Zealand.'[34]

The minister of agriculture wrote to Bob that his 'outstanding services to organic husbandry have been justly recognised with your award'.[35] His commendation gave official credence to the work Bob had undertaken for so long at Lincoln. Bob wrote in September 1997 that he had been to Wellington to receive his medal: 'Three of us wore *morning suits*, the only ones to do so but got commended by the Guv [governor-general] and his Aids de Camp for our turn out ... Boy did I look stunning.'[36]

Bob set off to the UK for the 5th IFOAM International Conference on Trade in Organic Products in Oxford in late September, to be followed by the IFOAM World Board meeting. He was looking forward to a trip to Devizes.[37] The trip started well enough despite the loss of his washbag: 'How little things like this can disturb me. Almost like an OMEN of worse to come.' Once in England he relished the 'mist & mellow fruitfulness. The horse chestnut outside the window gently stirs its autumnal russeting leaves and all is well with the world.' He approached the journey 'with a great thrill of anticipation. Almost as if for a trip of a lifetime. We shall see.'[38]

The conference began with an address from Peter Seger, an influential Welsh organic farmer, who Bob said conveyed the impression that 'there is immense opportunity with Organics set to explode'. The dinner, however, was 'soggy, no salad, bare tart slice and THEN they put out ash trays for a lack lustre after dinner speech'. This was not the vibrant organic world Bob had committed his life to creating. 'I went early to bed and felt *Depressed*. What am I doing here in this crowded, frantic environment. I need a sleep. But I awoke again at 4am despondent.'[39] A meeting he was called to with the World Trade Organization, about attempting to apply trade rules to organic labelling, seemed a 'lot of Gobble-dy-Guck!!' He 'left in despair. Really feel I am outside all this.'[40] To revive his spirits he took a walk in the meadows and along the river, noted the weather (light anticyclonic gloom), and had a drink with Ursula Soltysiak, an IFOAM World Board member from Poland.[41] The following morning he awoke again at 4am with 'thoughts/more thoughts about

Bob in Britain, 1997.
FROM TOP: *Cloud watching; in Martin Stokes' garden in Buxton; with Martin outside the Cat and Fiddle, the highest pub in England, in Peak District National Park; in Martin's garden, from left, Martin, Steven Craik, Bob and Martin's partner.*

the whole futility of the exercise. Should I resign? What am I here for?'[42] At least the final dinner – the Silver Jubilee – was excellent: Bob donned his Morris dancing outfit and joined the Oxford City Morris side for the entertainment.[43]

In London he met his old girlfriend Wendy, who was subdued following the death of her brother-in-law.[44] He then travelled to Sutton Courtney Manor House to continue the IFOAM World Board gathering. It was 'a lovely retreat, wild, rough kept but very atmospheric'.[45] He was more positive about this gathering: 'Meeting pretty good with good comment and decisions and I managed the p.m. chairmanship O.K. from 2pm–nearly 8pm. All quite amicable despite some pretty far-reaching decisions on Third World [management]!'[46] Further discussions at this event were 'good & meaningful as for Regionalisation others long and picky as for Internal organisation'.[47] The final day was:

> delightful … culminating in a row on the Thames which included Takele Teshame my favourite African from Ethiopia … Some interesting comment & debate and a general pleasant level of achievement. Terminated at close to 4pm in time to enjoy a lovely hour on the Thames with Bernward, Ursula, Ula!, Takele before a long talk, dinner at the George & Dragon and a final drink with everybody. Really good feelings.[48]

Although Bob did not relish the bureaucracy that necessarily underpinned the international organic movement he was so involved in developing, he felt they were making good progress. With time to connect with nature and enjoy the company of his co-creators, these fellow travellers, any doubts about his involvement were dispersed.

The rest of his trip proceeded as most had over the years: Bob visited many old friends, such as Martin Stokes, and explored the countryside. In Manchester he spent an evening at Via Fosse on Canal Street – 'the best gay pub ever I think. In fact the whole Canal Street area is a Gay Village … Quite amazing. Excellent evening,' wrote Bob.[49] He spent his final morning in Devizes walking at Roundway Hill, where he had scattered his father's ashes.[50]

Once home in Christchurch he noted, 'Murray did not visit.'[51] Madge had become 'a worry, very weak and fragile', and Bob reported in October 1997 that she had fallen, 'broke 4 ribs, stopped breathing, was resuscitated … looked like death warmed up and could hardly talk … How much longer will it go on, I wonder.' He wrote again in December that she was 'surviving. She is not coping with the situation which is a bit sad but understandable.'[52]

•

Following his retirement from lecturing, Bob commenced a contract with Lincoln to manage the upkeep of the BHU. One person from the organic community who provided enormous support for him at this time was Anne Seyger, now chair of the Soil & Health Canterbury branch. Anne was a committee member of the Canterbury Herb Society and organised monthly working bees with members to upgrade the unit's herb border.[53] Along with Sue Beauchamp and Nicole Bührs, she was one of several women who had helped regularly at the BHU for many years – sort of 'permanent Wwoofers', as Geoff Barnett described them.[54]

Anne was planning a twenty-first birthday party for the BHU to be held in April 1998, thus marking 1977 as the date of the unit's origin. The party would be a collaborative affair spread over a weekend with numerous activities and, of course, a banquet.[55] With Bob's flair for event planning, the 'celebration' blew out into a full-scale conference. The small organising committee – Bob, Anne and Charles Merfield – grew to include Holger Kahl (whom Bob had first met at IFOAM in 1984, and who was now teaching organic horticulture at Christchurch Polytechnic Institute of Technology (CPIT)), and Ray Wright and Gilda Otway from the Organic Garden City Trust. Holger sourced funds through CPIT to cover the cost of international speakers. The possibilities for using the event to further develop organics were manifold. The president of NASAA (Australia's organics certification body) would be invited to speak 'as a means to get an IFOAM regional group going' – an idea that had been on the table for many years.[56] The event would begin on the Friday afternoon with the Soil & Health Association's Annual General Meeting; Saturday's programme featured speakers on the topic 'Organics comes of age: Celebration of the past, vision for the future'.[57]

Mayor Vicki Buck was invited to open the 'convergence'. Bob included a personal note with her formal invitation: 'The BHU … does not appear to be a valued facility. So far it has survived the rigorous culling at the university but its future does still hang in the balance.' He explained that he had been contracted to the university for the past year to take care of the unit; one of his tasks had been to convene a BHU advisory group, which was working on a vision statement to be taken to the vice-chancellor in the new year. The statement was 'based on developing the BHU into a BIO-Village demonstrating appropriate technology as it relates to environmental self-sufficiency and sustainability based on appropriate resource use and in particular the use of the land for food production'.[58] He hoped that 'all these efforts will converge for the celebrations and perhaps form the basis for an official launch of the BHU towards a new tomorrow'. He acknowledged Vicki's work in establishing the Sustainable Cities Trust and the Organic Garden

City Trust and added, 'Support of Christchurch City through yourself could be quite crucial to the success of this venture.'[59]

Four months out from the event Bob expressed private doubts: 'I am organising a 21st birthday Conf. for the BHU in April but having second thoughts due to the uncertainties of survival,' he wrote in December.[60] OGCT trustee Ray Wright wrote in February 1998: 'Bob has put too much into the unit to just have it fade away – we all need to celebrate the impact of the unit and Bob Crowder on the Organic Movement in Canterbury.'[61] For his part, Bob was 'happy there should be a public BHU day but only to the extent it is a celebration of the BHU.'[62]

The event was a resounding success and drew even larger crowds than the record-breaking 1994 IFOAM conference. More than a thousand people turned up to the open day on 19 April 1998. Seven guides were kept busy with the crowds throughout the day. Among them were Trevor Jackson, who had studied Brussels sprouts at the site in the late 1970s and was now working at the Canterbury Science Centre, and Brendan Hoare, who had been elected president of the Soil & Health Association the day before. David Musgrave was based in the agricultural rotation where he shared his knowledge of the mixed herb ley, while Charles Merfield was situated in the amaranthus demonstrating flame weeding and tillage equipment.[63] Nicole Bührs, who had arrived as a volunteer in 1991, ran a tour of the whole area throughout the day. Vicki Buck was celebrated at the event: she had been awarded an honorary doctorate from Lincoln. Long-time BHU manager Geoff Barnett was also acknowledged: he was moving on to develop his own organic farm in nearby Motukarara. Geoff was now job-sharing with Helen Duckworth, who had become involved with the BHU in 1992.[64]

In the Soil & Health Association newsletter, Bob acknowledged in particular the encouragement of Holger Kahl and his team at CPIT.[65] 'We have much lobbying to do yet,' he remarked. 'But the flame burns brighter now thanks to the success of the open day and hope springs eternal.' Several Lincoln notables had attended the event. 'Hopefully they will see that the BHU has value that will equal its price and even allow for expansion into a NEW VISION for the future.'[66]

•

In July 1998 Bob set off on a six-week trip to Poland, Germany and the UK, partly for IFOAM business and partly for pleasure. Trips for IFOAM meetings had always afforded him an opportunity to visit family and friends and places of interest. He was now wrapping up his decade-long stint on the World Board. In Warsaw he and the team met with the Polish deputy minister in the Ministry of Food and

Food Economy, who 'appeared to be interested in what we had to say'.[67] Later they drank vodka with the deputy minister of agriculture.[68] Bob recorded that Bernward Geier was 'confessing to burnout', perhaps as a result of the 'apparent appalling disorganisation of Argentina 98' (the next IFOAM conference), and helped to draft an email to the Argentinian organising committee 'to try and remedy an almost hopeless situation'.[69]

In Poland Bob found 'rolling, wooded, floriferous country', a farm that was 'alive with insects, butterflies, *skylarks*', and an ecology centre that was 'really tranquil & idyllic'.[70] In Krakow, a 'wonderland', he discovered a café offering 90 types of coffee, and explored the castle, the cathedral and the Old Jewish Town. He enjoyed an ice-cold beer in the market square, attended an organ recital in the Jewish quarter, heard the trumpets signalling 11pm and midnight and visited the Białowieża primeval forest.[71] He went on rambling walks each morning before breakfast, one of which was particularly memorable: 'Wished I had my camera this morning. Farmer on traditional wagon 5 cows tethered to it, 2 horses and off to the fields – singing, stork among the haycocks to left and in the distance at end of cobbled road between the fields – the great Primeval Forest of Bialowieza. What an end to a delightful visit.'[72] The scene confirmed Bob's vision for a simple country life in biodiversity-rich fields with high culture accessible nearby. And it didn't hurt that there were handsome young border guards to chat to along the way.[73]

After Warsaw and a sobering visit to Trblinka, he travelled to Bonn to see his brother David and his son Simon, to celebrate Simon's birthday and inter David's wife Christel's ashes.[74] In Britain he visited the Lost Gardens of Heligan, Elm Farm, Ludlow (to see former Wwoofers Richard and Bijon) and Penzance. He visited Wendy in the Scilly Isles and climbed Kinder with Martin Stokes: '10 mls or 15–16kms plus a climb to 2000 feet. Needed that cup of tea & cake ... A lovely day.'[75] He caught up with Steven Craik and a meeting of the Gateshead Gay Group, and together he and Steven drove to Edinburgh.[76] There he visited former Wwoofers Stuart Jeffrey and Page Lawson and delivered a paper at a symposium on organics and sustainable permanent development: '*boring* papers including key ones ... My own went well as it happened BUT felt *used* – just a thank you, no offer of lunch dinner or even a *drink*! ... What a *bore*.' Steven collected him and they went to a coffee house where an evangelical group gave them free coffee and cakes: 'what contrast'.[77] He went to Sherwood Forest, reacquainted himself with Nottingham and visited his old rowing club's new home at Holme Pierrepont National Watersports Centre.[78]

On 17 August Bob returned to New Zealand and the ongoing debate about the future of the BHU.

Dereliction: the intensive beds established originally by Tim Maples and tunnel house now a riot of unproductive wilderness, November 1999.

9: THE DEATH OF THE BHU?
1998–2000

SOIL & HEALTH CANTERBURY continued to hold events and working bees at the BHU throughout 1998, even while Bob was overseas during the winter.[1]

The work involved in organising the open day in April had brought the OGCT to the table, and in July that year the trust began to make its views on the BHU known to Lincoln University. Bill Kain, director of the Postgraduate and Research School at Lincoln, responded on behalf of the vice-chancellor: 'I appreciate the immeasurable value the BHU has for the wider organic community in New Zealand ... [T]here is considerable benefit in having organisations like the OGCT involved in setting the future direction of the BHU and [Lincoln] would welcome your involvement.'[2] He stated Lincoln's interest in developing relevant undergraduate and postgraduate degrees in organic production and in ensuring that the directions and operations of the BHU supported that objective through enhanced links with the organic movement and the investigation of national and international initiatives for cooperative education. The OGCT shared a Statement of Strategic Intent with Lincoln in August 1998, and a meeting between Lincoln and the OGCT management team was planned for 10 September.[3]

Brendan Hoare (who, with Bob's backing, had become president of Soil & Health in 1998, and who was a former manager of the BHU) was determined to get the BHU back on Lincoln's agenda. According to Jon Manhire, who at the time was executive director of the Organic Products Exporters of New Zealand, Brendan brought the relevant parties together and 'was probably the greatest driver for the re-establishment of the BHU'.[4] Brendan was clear about his motives:

> You can't ... have a movement unless you've got an educational base ... So the destruction of the BHU to me was sacrilege ... We needed that base ... The BHU was a seeding ground ... So I put full weight in behind that because it was strategic ... How dare they destroy that space? ... Saving the BHU for me was like a mission ... What [Bob] had created was bigger than him. Way bigger ... I was so happy to be able to do that, firstly for the BHU and the role that it had in society and in New Zealand, but also for my mate Bob.[5]

As a result of this activity, a meeting about the future of the BHU was held at Lincoln on 1 October 1998, attended by Bob, Holger Kahl (OGCT and CPIT), Joep Nederpelt (chair, OGCT), Ian Cornforth (ex MAF, now Lincoln), Bill Kain

(Lincoln), Seager Mason (BioGro), Brendan Hoare, Rex Verity (Christchurch Small Business Enterprise Centre), Charles Merfield and Ian Spellerberg (Lincoln).[6] It was a (male) 'who's who' of the New Zealand organic movement, as well as key Lincoln staff.

Ian Cornforth 'was keen to see the science focused and on track' and bluntly told the meeting that 'any decisions taken would not negatively impact on his budgets, which the BHU falls under'. Brendan asked what Lincoln's philosophy and business direction was, in order to understand its apparent new-found enthusiasm for organics. Ian replied that his 'focus lay in large-scale organics and environmental production systems as that is where funding could be secured'. Ian may have been encouraged by Seager Mason's report that New Zealand's organic export market – worth $40m in 1997 – was projected to climb to $100m in 1999, along with 25%+ rates of growth in the US, Europe and Japan, and that these markets were consumer- and market-led. Brendan reminded the meeting that Soil & Health had invested $51,000 in the BHU over 15 years.[7]

The meeting resolved that the BHU should be maintained until a business case was developed, to be completed by 1999. The unit would be the resource base for a summer school on organics in 1999, and links between the BHU and Lincoln's degree programmes should be developed in terms of resource sharing. The BHU's Wwoof offering had ceased to exist shortly after IFOAM '94, but now accommodation would again be made available for Wwoofers – a recognition of their important contribution to the upkeep of the unit. Lincoln hired Jon Manhire to develop the business plan (it was completed in 2002).[8]

This work led to a draft Memorandum of Understanding regarding the BHU between Lincoln University and representatives of the organic movement, dated 11 November 1998 and further developed by 23 April 1999 into an agreement with 21 clauses.[9] These included Lincoln's practical and philosophical commitment to organics: that organics be integrated across the curriculum, that organics qualifications be developed, and that quality research in organics be promoted and encouraged. A full-time academic appointment for a 'qualified and experienced proponent of organics' would be made, and a full-time BHU manager 'with organic expertise and people-management skills' would be employed. In addition, an Organics Advisory Board made up of signatories to the MoU would be established and this, with Lincoln, would form 'an Organics Trust to help fund organics education, demonstration and research'.[10] For now, the MoU seemed resolved.

No practical help was forthcoming in the meantime, however, and the situation was becoming unsustainable. On 21 September 1998 Bob informed members of Soil & Health Canterbury and the Good Gardeners that, with Geoff gone, 'there is no one working at the BHU now ... they are just managing or maintaining it and taking it on a day by day basis.'[11] Agroecology student Charles Merfield, who was also involved with Soil & Health, CCOG and the OGCT, wrote that he saw a significant opportunity ahead for the BHU, since plenty of academics and managers at Lincoln viewed organics as part of the future of agriculture: 'There are people at Lincoln who consider such things to be attainable, but only if the organic community physically comes together with coherent proposals and demands.'[12]

Bob did not agree. 'With all due respect to Charles Merfield,' he wrote, 'the answer does not lie with the organic movement but rather within the university itself. After all, without the support of the Organic Movement over the years the BHU would not be here.' He reasoned that, at heart, the BHU stood in opposition to the university's own values; it would either be rejected wholesale or reconfigured to fit those values. 'The BHU is a centre of inspiration, teaching, demonstration and research devoted to a vision of sustainability that exists outside the existing paradigm, putting environmental reality ahead of a flawed economic reality.' The apparent rejection of the unit highlighted the many issues pertaining to the neo-liberal economic paradigm. 'This is difficult for scientists and economists to take. Despite obvious indications from the environment, and the increasing power of aware individuals, there is a reluctance to accept a pathway that is such a paradigm shift.'[13]

Anne Seyger remained optimistic. She reported in February 1999, 'The management issue for the BHU has not been resolved yet, but a dedicated team of Wwoofers guided by Bob Crowder is working wonders ... the BHU is still alive and kicking.'[14] She reiterated her message in May, adding, 'No official verdict has been made on the future of the BHU.'[15]

Anne may have been patient in her commentary, but not everyone was. Shen Khen Heng, who had been involved with the BHU as a student for three years, blasted Lincoln for its lack of attention. 'It is unbelievable how [the university] has been able to so blatantly lie to the world about how wonderfully environmental they are, with glossy printouts of their "Environmental Policy" that anyone with half a brain that bothered to look could tell was just rubbish.' Shen considered that the stated desire – to weave organics and sustainable farming into the curriculum – was 'no more than ten minutes of error-ridden and misguided criticism of organics by red-necked lecturers'. The BHU was 'like a rose bush growing in a pile of horse crap.'[16]

Shen's anger was understandable. Bob himself reported that since the open day on 19 April 1998, at which Geoff Barnett's retirement had been notified, the university had completely withdrawn 'finances and goodwill'.[17]

In March 1999 Bob received a message from Jamie Trachta-Dyson in County Wexford, Ireland, inviting him to offer advice on the establishment of an organic orchard owned by well-known philanthropist Michael Watt. Jamie was tasked with setting up the orchard and a vegetable operation and had read about the use of umbelliferous understories in Nic Lampkin's *Organic Farming*, which featured images from the BHU.[18] Bob's reply acknowledged the importance of soil health and compost and 'the diversity of the surroundings because it is there that the balance in the fauna takes place. That is why you must build a stable environment through diversity in the flora that will encourage that balance in the fauna, especially the insects but not forgetting the birds who will also respond.'[19]

Born in Christchurch, Michael Watt made his fortune in television. In New Zealand he is probably best known for his support of cricket, in which he invested at least $2m, including in the Bert Sutcliffe Ground at Lincoln University in 2000.[20] After Jamie's initial contact, Michael expressed his interest in the BHU and invited Bob to visit him.[21] Bob was planning to travel to the UK in May to walk Offa's Dyke, a 285-kilometre trail on the border between England and Wales, as a sixtieth birthday present to himself. He would be able to combine this with a visit to Ireland. He phoned Michael and told 'him the story of the BHU and the indifference of Lincoln University towards its value and achievements ... which has culminated in the complete withdrawal of funding and goodwill since 19th April 1998'. The pair got on well, and Michael requested a copy of Bob's resumé and some budgetary information on the BHU.[22]

Immediately after this promising call Bob drove to Lincoln, where he and colleagues agreed to send Michael the draft Memorandum of Understanding that had been drawn up the month before. They also sketched out a three-year budget. On 25 May 1999, one week later, Lincoln University development coordinator Ivo Wynn-Williams faxed these to Michael with the intriguing subject line: 'Lincoln University Centre for Organic Agriculture'.[23] Such a centre may have been mooted or implied through the discussions about the future of the BHU with the organic sector, but in this context it was perhaps an unhelpful attempt to describe the direction Lincoln was going. (There is some logic in the use of this name, however, and perhaps this had escaped the notice of the organic sector. In early May, Lincoln University advised the media that it intended to establish at least nine new research centres. One of these was a 'Centre for Organic Production' – a joint project between Heinz Wattie's Ltd, the organic movement and the university.)[24]

Ivo Wynn-Williams told Michael he was 'sure that Bob has told you that Lincoln is keen to expand its activities relating to organic agriculture and, as a first step, has prepared a Memorandum of Understanding between the university and members of the New Zealand organic movement'. As Bob was visiting soon, he would 'be able to give … a more detailed account of what we are trying to do'.²⁵

The budgeted expenses came to approximately $1m. Michael Watt replied the same day: 'You can take it from me that I am interested in the project and that the total operating expenses are within the parameters of where I would be coming from … before proceeding any further I would like to meet up with Bob Crowder as soon as he gets to England'.²⁶

Bob left for the UK the following day, 26 May, and waited for a date for the meeting to be confirmed. While he waited, another milestone passed: 'Last pay cheque today – L.U. contract ended'.²⁷

The two met in London, where Bob showed Michael a series of slides and talked through the challenges and opportunities.²⁸ Bob described the meeting:

> He stood up, walked around his desk and stood holding out his hand to shake. 'Congratulations,' he said, 'I think you have got yourself a new contract' … I stood there, stomach churning, a MILLION dollars over 3 years? The BHU saved, more money than one could dream of. A new lecturer in Organic Philosophy and Practice, a manager for the BHU itself and even a couple of associated workers to get the place back into shape for spring.²⁹

The money would enable Bob to achieve his vision and conclude his career on a high note. He communicated the exciting news to colleagues at Lincoln.

Michael invited him to visit his property in Ireland. The day before Bob arrived, Lincoln's Vice-Chancellor Frank Wood sent Michael a message of gratitude for the generosity the latter had shown 'to support the development of Lincoln University's Organic Centre': 'The Organics [sic] Centre promises to make a significant contribution not only to Lincoln University's developments in education and research … but to national and international producers and consumers generally.' He continued: 'Lincoln is making very good progress and has reaffirmed a strong commitment to supporting New Zealand's advancement as a world leading bio-economy. Your sponsorship will contribute significantly to our support for a very special and growing area of opportunity.'³⁰ Frank Wood's message mentioned neither Bob Crowder nor the BHU.

On 10 July Bob arrived and was hosted by Michael and his wife Gaynor and their children, and two days later they finally managed to have a proper discussion. Bob felt it was '*Difficult* to get whole story across. Finished about 11.40pm.³¹

Exactly when Michael received the fax from Frank Wood is not clear, but he confronted Bob about it on 13 July. He appeared furious and said that if what Lincoln was planning was a commercial undertaking, then it should be financed by the commercial sector.³² Bob described the moment: 'Mike brings letter from

V.C. L.U. & asks for comments. Asks me to draft a letter for him to V.C. Problems looming.'[33]

Bob was also disturbed by an email from Charles Merfield.[34] This was likely one that Charles sent to the MoU signatories group on 8 July, in which he informed the others that the version of the MoU Ivo Wynn-Williams had sent to Michael Watt was different from the one agreed to – after months of tortuous discussion – between Lincoln and the organic movement. Frank Wood, who had received the final draft in April, had made substantive changes, and the revised version had been sent to Michael Watt before the co-signatories in the organic movement had seen it.[35] Charles wrote, 'Lincoln has … acted in very bad faith and considerable arrogance, in that we negotiated the previous final draft over several repetitions with a senior representative from Lincoln … The changes … considerably weaken the place of organics at Lincoln and our original thrust.'[36]

Bob was now in a bind. At Michael's request, he drafted a response to send to Frank Wood. It was short and to the point: 'My interest in Lincoln University, as associated with the present move to organic systems, is entirely associated with the work of Bob Crowder over the last 20 years in the establishment of the Biological Husbandry Unit.' It stated, 'Lincoln University had … no resources to maintain the unit, and … it was already in terminal decline due to the withdrawal of funds by the University.' The letter commented on the MoU – and here Bob was treading a delicate line: 'I have read the Memorandum of Understanding of 25th May 1999, between the University and the representatives of the NZ Organic Movement,' he had Michael say, 'and see that there is a will to continue the work of the BHU. The aim is to continue BHU's position as a self-sufficient, sustainable centre for teaching, demonstration and research. This is the concept that I want to support.'[37] It appears Michael smelled trouble and never sent the letter.

Two problems needed a coherent and immediate response. The first was about the term 'Organic Centre': the issue that concerned Michael. Charles explained in an email on 15 July that 'Ivo was the one responsible for the phrase "organic centre" rather than "BHU" and claims that it was a mistake on his part and that the intent has always been that Mike's money will go to the BHU and academic work, not into some general organic fund. I believe that this is correct as Ivo's file has a list of key points for an agreement between Lincoln and Mike and they are about the BHU not an "organic centre".'[38]

The second was about the revised MoU, which had so dismayed the organic movement's signatories. As Charles commented, 'What is not clear is if Bob has seen Frank's MoU. The only way he could have seen it is if he saw the copy that Ivo and Steve [Wratten] sent to Mike.'[39] Bob had indeed seen the altered version.

Brendan Hoare, Bob and Haikai Tane, Auckland Airport, 1999.

Brendan Hoare called on everyone to keep 'cool heads'. 'Bob is only a few days away. We wait, neither confirm or deny, allow Bob to handle the man with the donation in the UK and be patient. Do not be brash, or quick to make decisions until we know the facts.'[40] Unfortunately, the fiasco had done irreparable damage.

On his return to New Zealand Bob undertook to rectify the misunderstanding. He retained high hopes that a deal could be struck. He was met at the airport by Brendan Hoare and Haikai Tane and the trio celebrated the success with champagne. But Michael Watt had lost interest. The deal was off.

How could a revised MoU have been sent to Michael before the signatories had agreed to it? Bill Kain wrote to Bob, apologising sincerely 'for any embarrassment and loss of trust that the process … may have caused you'. There was nothing sinister to be read into what had happened; he took full responsibility. 'Unfortunately, when I left to visit research organisations and universities in Europe and the UK, I did not leave clear instructions outlining the appropriate actions that needed to be followed to ratify the revised version of the MOU before presentation to a third party.'⁴¹ But the fact that the decision to send Frank Wood's version rather than the ratified version had been made by Ivo Wynn-Williams and Professor Steve Wratten rankled Bob and made him wonder: had Steve tried to manoeuvre money into his own project, Kowhai Farm, Lincoln's joint venture with Heinz Wattie's?

In early September Bob wrote a letter to Michael that reads as a final communication. He confessed that he knew the MoU Michael had read was not the one the organic movement and Lincoln had agreed to in April:

> Needless to say I do feel a bit let down by the eventual non-event resulting from our all too brief discussions. Still I appreciate your time and the brief flicker of visionary fervour I experienced as a result of our meetings.
>
> For myself I need to consider life without Lincoln and today I have finally cut the umbilical cord of over 30 years.
>
> The bitterness is there and I believe a great opportunity lost.⁴²

Bob's letter also explained that the mixed-cropping organic farm, the 57-hectare Kowhai Farm, had been launched the previous week: 'I support the move even though I had to hear about that venture via rumour and gossip.'⁴³ The launch date was 26 August 1999 – a month after Bob's return.⁴⁴ According to *Organic Matters*, Kowhai Farm was a 'commercial organic demonstration farm producing crops and livestock'; it was 'to become a showpiece for the burgeoning organic industry in New Zealand and is expected to attract international interest'. The article quoted Frank Wood: 'Organics in New Zealand will take on a new perspective as the initiative builds on pioneering work carried out at the Biological Husbandry Unit.'⁴⁵

The Press reported, 'Many people concerned about genetically modified foods will welcome the organic-farming project planned by Lincoln University and Heinz-Wattie's … Its promoters hope it will encourage more farmers to change to organic methods.'⁴⁶ In a letter to the editor, Bob retorted: 'As your editorial … says, many people will welcome the arrival of organic farming into Lincoln University farm land. Others might wonder what has taken so long?' He recalled the 20 years of organic philosophy and practice at the BHU and said it was 'depressing therefore to tell your readers that Lincoln University has consigned those decades of experience to terminal decline by the withdrawal of all financial support for the unit. The BHU together with its inspirational aspects of teaching, demonstrations, and research are dead.'⁴⁷

The effective closure of the BHU in 1998 and the opening of Kowhai Farm in 1999 was a clear statement from the university about its planned role in the development of the New Zealand organic sector. Worse, immediately after Kowhai

Perfect broccoli in the last year of semi intensive rotation, c.1996.

Farm opened in August, Lincoln made the drastic decision in September to clear 'all seed lines and records ... from the one remaining [BHU] office with no prior notice'.[48] It was a particularly harsh blow for Bob: a kind of attempted erasure, perhaps, and maybe even a declaration of war. 'It is plain that the BHU is well outside the existing paradigm supported by Lincoln University and therefore must be cast out in a great "cultural revolution"', he wrote.[49] 'The University has at last closed down the Biological Husbandry Unit and will re-invent the organic wheel in its own image, with no reference to the past and least of all to Bob Crowder.'[50] Brendan Hoare recalled that Bob was 'angry ... he was disgruntled and down ... he felt ... disowned.'[51]

•

Ironically, it seemed the organic way was finally getting somewhere after the slow progress of much of the 1990s. Nationally, among other changes, an Organic Products Exporters Group was established at the end of 1995; locally, the OGCT had developed in Christchurch in 1997, and now Kowhai Farm was underway as well.[52]

Organics education also received a fillip. In 1998, as the BHU was shutting down, Brendan Hoare began work as a lecturer in sustainable production systems at Unitec in Auckland. He established the Unitec Hortecology Sanctuary Mahi Whenua, a diverse and abundant food garden that echoed elements of his beloved BHU and incorporated ideas he had gleaned from Haikai Tane, Stuart Hill and his studies at Queensland University of Technology. Thanks to Brendan's influence, Unitec hosted 600 people at the Soil & Health conference in 1999.[53] In the same year, Brendan spearheaded the Organic Federation of New Zealand (OFoNZ), which comprised Soil & Health, BioGro, Demeter and the Organic Products Exporters Group.[54] In his words, the establishment of OFoNZ required the smoking of many peace pipes between parties.[55] It was the peak body for organics in New Zealand and a direct point of contact between the sector and the government. In time it became Organics Aotearoa New Zealand (OANZ).

Brendan also worked on the Green Party's Organic Eco-Nation policy and used Soil & Health's magazine (which he had renamed *Organic NZ*) to promote that vision as much as possible.[56] The decade ended with the Green Party getting into Parliament in its own right; with it came an agricultural policy that presaged positive change across the sector.

Politics might not be the only answer, of course. 'The challenge now is not to become even more complacent with the comfortable belief that now they're in

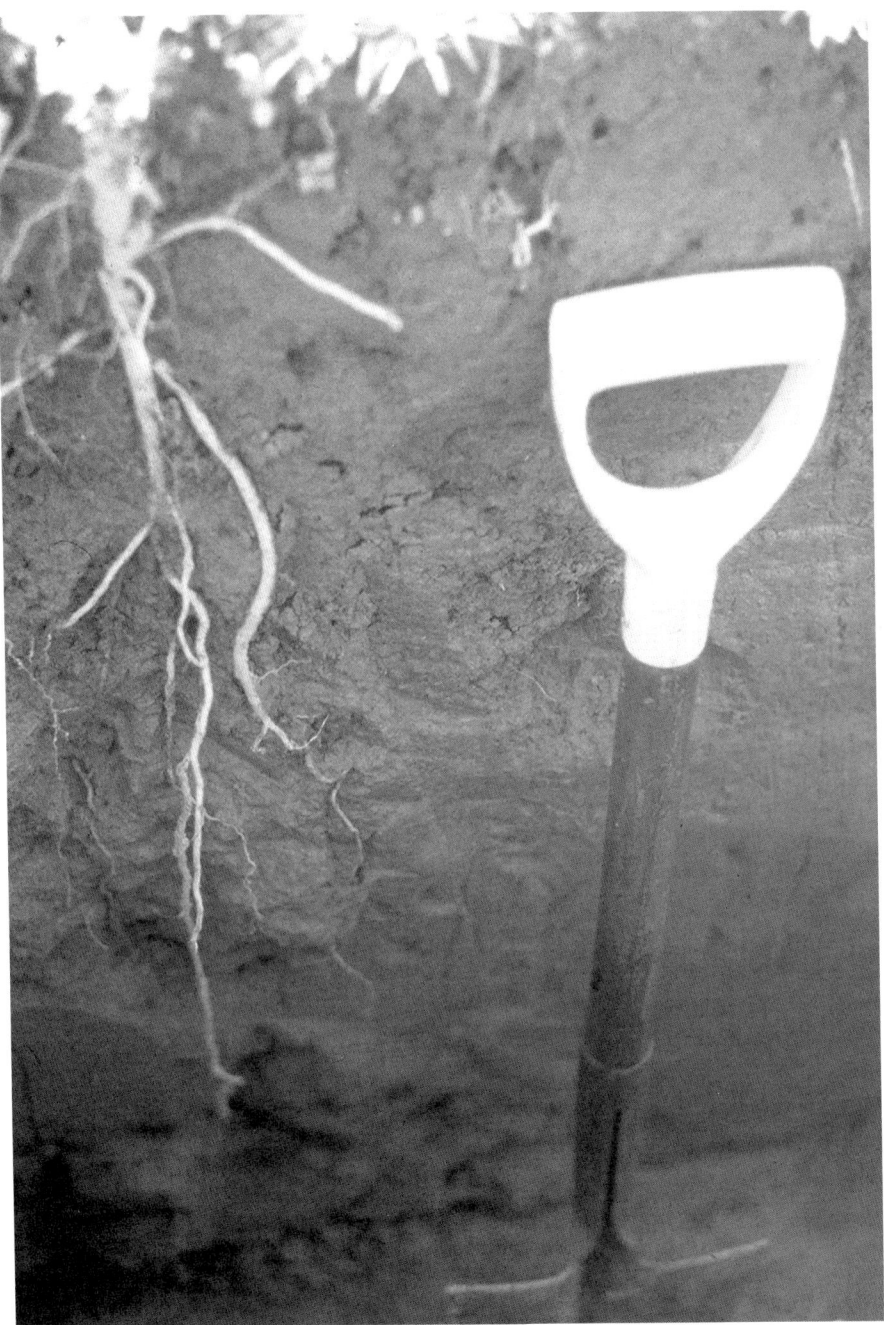

Chicory and lucerne roots incorporated into a mixed herb ley, penetrating through a soil pan to an incredible depth at the BHU. This built soil structure and allowed water infiltration.

Parliament and brilliantly represented, we don't need to do anything ... With the BHU wrangling ... we get a taste of the damage ... politics can do to the movement as a whole.'[57] Yet politics could help: after all, the Green Party was elected on the back of a campaign centred on a popular anti-genetic engineering message. In 2000, Jeanette Fitzsimons hosted a meeting titled 'Organic NZ by the year 2020', co-chaired by Brendan Hoare and Chris May, where the question of how to achieve an organic vision for the country was fleshed out.[58] Plans were hatching for government funds to be released towards the organic sector, and the Soil & Health Association, under Brendan's leadership, held a high-profile conference in Auckland's Aotea Centre at which it launched its vision for the future: 'Organics 2020'.

Work was also underway to create the BHU Organics Trust, as flagged in the MoU between Lincoln and the organics movement. The organic community was involved but sceptical: 'It remains to be seen whether words can now be put into deeds', the OGCT commented.[59] The trust was launched on 11 August 2000. Robyn Patchett, of CCOG, reported that it 'was a great celebration for many organic people from far and wide, for Bob Crowder, for the many volunteer workers, for so many distinguished people that Vice-Chancellor Frank Wood didn't individually name them, and of course Lincoln University!'[60] Frank Wood thanked Bob and the team who had created the trust and noted the huge potential for organic exports and Lincoln's ability to support this with research; Bob's reply focused on the students who had so energetically influenced and supported the vision in the 1970s. His speech, reported in the COGS newsletter, illustrates his thinking at this time: his commitment to philosophical principles that included recognition of the people who had supported the work with quiet determination:

> Mr Crowder emphasised that biological husbandry must have a philosophical base, there must be somewhere where the concepts of organic philosophy are nurtured – caring for the environment using organic production methods, working towards a total health, in other words based on holistic principles. He saw the B.H.U. as involved at this level. The mainstay of the unit over the years was Geoff Barnett who was responsible for the day to day running. Early Organic growers – Tony Mallard, Ernst Frei and Tim Chamberlain – gave their time and expertise, Holger Kahl of the Christchurch Polytechnic had supported the unit and many Polytechnic students had worked at Lincoln. Last but not least Mr Crowder thanked all the voluntary workers who had given time so generously – Helen Duckworth, Nicole Bhurs [sic], Anne Seygar [sic] and many more. His final quip was that 'the unit has not had any work done on it for two years now and it's still there – testimony to the durability and sustainability of organic production.'[61]

The chair of the BHU Organics Trust was Sir Peter Elworthy, who Peter Waugh recalled as an enlightened farmer politician. Sir Peter described his organic influences, among them Bob, the IFOAM conference and 'the obvious demands of the market place':

> On his travels he had seen an 'unstoppable, escalating demand for safe traceable organically produced foods'. Sir Elworthy saw the BHU as a resource and educator for not only those involved in organics but the universities and the CRIs [Crown Research Institutes]. The skills needed for organic production are greater than for conventional, there is also a need to be more flexible, and a need for better planning. In Europe there were subsidies to help get started in organic production; here farmers have difficulty seeing the benefits and remain unconvinced of the market advantages. There is a great need for new technologies and extension work. The establishment of the trust represents a new stage and the calibre of the members of the trust will ensure its success.[62]

Soil & Health Canterbury's president Peter Green afterwards reflected:

> Lincoln University appears fettered by the need for huge international funding and our government's insistence on a bio-technology-based knowledge economy. The Biological Husbandry Unit has been re-launched, but they have yet to discover the … huge resource that awaits them in the passionate energies of people in our bio-region's organic community that were so much a part of its internationally acclaimed success in the past and could be again.[63]

Things went awry with the new trust almost immediately. Soon after the official launch, at which Bob had memorably presented Frank Wood with an olive branch from his own garden, Sir Peter called a meeting to make certain decisions regarding works on the BHU site. Bob later contended that he was unaware of the meeting and was upset that it took place while he was on a lecture tour in Australia.

Sir Peter was apologetic and tried to explain the situation as a misunderstanding. Decisions made at the meeting, which largely concerned orchard pruning, were, he said, to be confirmed at another meeting on 8 September at which Bob would be present. Bob did not accept Sir Peter's explanation and pointed out that the minutes of the unscheduled meeting gave no indication of decisions being deferred. In fact, he wrote, 'on the Friday following my return Tony Whatman rang me to inform that he had organised the commercial orchardists for the coming Monday and would I be available to advise them on what was wanted on the ORGANIC ORCHARD because it contained varieties they, the commercial pruners, were not familiar with!!'[64] He concluded, '[It] appears to me that this whole issue is very similar to how I have been marginalised by Lincoln in the past to the serious detriment of the welfare of the BHU. I gave Lincoln an OLIVE branch but have received a gooseberry in reply and I can see no future for myself on the trust. I just can't work that way.'[65]

Sir Peter and Bob knew each other well – the Morris dancers performed at the annual daffodil festival held at the Elworthy's property, Craigmore. On the morning of the follow-up meeting on 8 September, Sir Peter and Lady Fiona Elworthy visited Bob at home.[66] He was out roaming the Cashmere Hills and checking his rain gauges, and although they waited for him for some time, they did not see him.

Soil & Health Canterbury branch wrote to Sir Peter, wanting to know 'how it was that certain major decisions could have been made during a meeting that had not been properly notified while a key member of the board was overseas', and recording its disappointment that Bob was minuted as having given an apology for the meeting 'when he apparently had no idea the meeting was taking place'. The committee found it 'disconcerting that the provisional nature of these decisions were not adequately minuted'.

> The BHU has always been not merely *the* practical demonstration unit for holistic organic philosophy par excellence; it has also been a demonstrable galvanising force for the organics community at large. Volunteers have played a large part in the unit's success from an economic point of view. They have also added a unique element of co-operation between the Establishment and the community, and this has been a vital aspect of the BHU's new paradigm approach. That the community has felt it has some 'ownership' of the BHU has been one of the principal reasons for its popularity throughout the country.
>
> We were surprised therefore to hear that a decision (revocable or not) had been made to contract pruning out. While we understand the need to get the job done quickly it does need to be pointed out that the prunings have generally been undertaken with assistance from the community, and we had hoped that Soil & Health Canterbury might have been approached in regard to this first.
>
> We recognise that the decision to contract the job out was revoked after hearing Bob's views, but it does seem to illustrate the point that the aspirations of local groups are not necessarily represented on the trust. If Bob is not to be present on the board, we are concerned that the problem will be exacerbated. We believe that a good feedback loop between the BHU Organics Trust and the community will help if the trust wishes to maintain the excellent links with the community that have been established over the years.[67]

The BHU Organics Trust Board met a few days prior to receipt of this email. Board members Mollie Chalklen and Ian Henderson were thanked for their work 'to establish a resolution in regards to Bob Crowder', although exactly what that meant was not explained. They had evidently tried to encourage Bob to stay with

Mixed shelter and coppice at the BHU starting to become unmanaged wilderness, 1998.

BHU, October 1999.

Closed for business. Poplars, established as part of Bob's original plan, felled across the BHU entrance, December 1999.

the trust. Acknowledging 'delays resulting from the Bob Crowder situation', the board agreed 'that a BHU manager would not start until next year, which would be too late to initiate much for this season'.[68] There was frustration on all sides.

Bob felt unable to rejoin the process. On the day of the board meeting he left town to walk the Banks Peninsula Track with friends to check the condition of the trail. Entries in Bob's diary indicate how much pleasure he derived from the company and the environment. On the first night the group enjoyed a sing-along, and the following day Bob delighted in Hugh Wilson's ability to point out rare ferns and other botanical highlights: 'the yellow corokia on the wind-clipped bushes on the exposed ridge', an 'amazing gorse display' that 'contrasted with the lush descent into the stream valley', the white clematis and 'amazing perfume' of the yellow, pretty waterfalls, mistletoes, and white-flippered penguins 'in their nests & at play in the bay'. At Ōtanerito Bob enjoyed a paddle in the sea followed by champagne cocktails.[69] Such experiences lifted his spirits and reminded him of earlier journeys with Hugh.[70]

In November 2000 the Organic Garden City Trust reported that Bob had officially retired from Lincoln.[71] It was a polite way of expressing his complete rejection of a process that had side-lined his life's work.

The BHU Organics Trust now entered difficult territory. Anne Seyger returned to Europe, the working bees ceased and the BHU fast became an unmanaged wilderness. The university blocked access to the unit by having poplars cut down and laid across the driveway. Bob visited from time to time 'to keep an eye on the decline of it because it got more and more overgrown ...'[72] The trustees sought proposals for interim uses. It appeared there might be some flexibility from Lincoln University as to what these could be. Bill Kain told the board that the Statement of Intent and Memorandum of Understanding signed by the various parties 'should not overrule questions of function. They are very prescriptive because they were developed in an atmosphere of mistrust, and the new level of trust on both sides will allow greater flexibility.' He continued: 'When the trustees know where the BHU should go, they could expect Lincoln University support in data collection and evaluation, education, training, and marketing questions provided that a variety of people are approached. One person cannot take the entire load.'[73]

At least that had been acknowledged at last.

Coffee morning at Bob's with Hans Schaper, Jerome Borel, Bob, Jeremy Ironside, Geoff Barnett and Lily White, 2021.

10: Retirement, 2001–23

PROGRESS WITH DEVELOPING THE BHU ORGANICS TRUST continued throughout 2001, and the trust's deed was formally lodged in 2002.[1] The listed trustees were Sir Peter and Lady Fiona Elworthy, Mollie Chalklen, Haikai Tane, Morgan Williams (formerly with MAF and now parliamentary commissioner for the environment), David Musgrave and Jon Manhire.[2] Most had been on the planning committee for IFOAM '94 and all understood the tremendous mana of the BHU.

In 2001 the trust selected Dr Tim Jenkins, one of Bob's former students, as manager of the BHU. He came armed with excellent skills, a strong research background, and, as Jon Manhire said, 'a lot of passion … he was actually the perfect person for that role'.[3] Tim began the task of bringing the wilderness back from the brink.

Tim had studied horticultural science with Bob in 1986 and recalled 'memorable lectures on weather, precision planting and of course a strong component of organics'.[4] From late 1988 Tim was part of the organic research team at Lincoln MAF, focused on biological control and plant pathology: '[I]t seemed like a flourishing time for organic research but this was comparatively short-lived – just a bit too early for the appetite of many in the prevailing economic climate and cuts in research funding.'[5] He visited the BHU regularly in 1989 on some joint MAF and BHU projects. In 1990 he returned to the University of Canterbury for his final year in a Bachelor of Science and was able to include in that a postgraduate course in biological husbandry, which was led by Bob at Lincoln. It was the first year the course had been offered:

> One of the components … was training as an auditor for organics … We did farm visits and Bob taught us the things to look for when auditing. He commented that what the farmers provided for lunch in the kitchen could be a clue as to how seriously they took organics.[6]

Tim then took up a position at DSIR Grasslands in Lincoln and Tekapo; he visited the BHU occasionally and prepared to do a PhD in biological control of weeds.

Tim's work as manager of the BHU was unveiled on 16 March 2002 on what Sir Peter called 'a truly historic day': the first public open day 'since the rejuvenation of the Biological Husbandry Unit … was begun six months ago'.[7] It was a modest event with some 30 people present. Sir Peter highlighted his own commitment to the organic community and emphasised: 'We want to not only tell you what's going on, but to hear what you think should be going on. We hope that this will be a two-way participatory process.'[8]

Mollie Chalklen, who had been so instrumental in shifting the focus of the organics fraternity from gardening to farming in the 1980s and bringing Bob firmly into the movement's orbit, also spoke. 'We're building on what Bob created,' she said,

> and we are going further because we have the land, the money, and people in high places who can talk for us, something which Bob didn't have. We hope all our Open Days will be a celebration of what Bob started, and what Tim, who is after all Bob's disciple, is able to carry on ... What we are doing here will be known throughout the world, it is a tribute to Bob, and a tribute to Tim. We've got the opportunity to make it something very, very special by taking the best of every system there is. I think this is very, very exciting.[9]

Mollie's claim that the trust board 'had the land' referred to the imminent signing of a lease between the trust and Lincoln University. It was signed in April 2002 and would expire on 31 December 2021.[10] The move signalled a much-improved relationship with the university and gave the trustees the certainty and confidence they needed to make progress.[11]

Tim Jenkins' vision for the BHU was demonstrated more extensively at a second open day in September 2002, when Mollie announced that some blocks of the unit had been renamed, to memorialise Eve Balfour, Albert Howard, Rudolf Steiner and Guy Chapman. The trust had also discussed using the names 'Crowder Heritage Block', 'Kyusei' and 'Pacific Polycultures' for an additional three blocks. Geoff Barnett and Brendan Hoare, neither of whom were board members, were considered the best people to approach Bob about these possible names – an indication of the fragile connection between the board and Bob.[12] A Crowder Block, along with a Maples Block, was eventually established.

Two innovations were to take place at the unit: the introduction of Japanese 'nature farming' with effective microorganisms (EM), and the introduction of sheep into the rotation. Tim's vision now for the BHU was a kind of 'fusion farming' to further research approaches that could then be applied on a larger scale at the Heinz-Wattie's Kowhai Farm. 'Both the BHU and Kowhai Farm are looking at resilient organic agriculture systems,' he said; the two areas complemented one another. 'Together we place Lincoln University right at the forefront of international research [on] organics.'[13] As Tim reflected later:

It soon evolved into a major focus on supporting organic small farms. Namely, the development and promotion of techniques for making organic small farming more successful and providing an incubator and information for small farm growers. I think this was the most important goal and output for the BHU and a way to make 'model systems' more sustainable (in terms of being able to keep running them without requirement for direct funding) and more relevant (it would help achieve the aim of more people and more success in organic small farming).[14]

Tim's six-monthly report of August 2003 summarised an impressive range of activities and improvements and noted that the BHU now had three full-time and one part-time staff member as well as a number of volunteers.[15] The unit was providing a field-study site for numerous master's and PhD students and had synchronised well with Steve Wratten's research programme. Tim recalled, 'Steve had an extremely high international profile in Ecosystem Services, the idea that one could grow beneficial flowers or habitat species ("beetle banks" or "spider strips") to improve the levels of biological control in crops. We benefited from this association and promoted the concepts regularly and in the written materials we produced.'[16]

One of those research projects was conducted by Charles Merfield, who, since his arrival at Lincoln in 1994, had completed his Master of Applied Science.[17] Throughout this period he worked almost continuously with Tim Chamberlain. In 2002 he commenced his PhD at the BHU. His topic was organic carrot seed production, including organic controls for fungal disease, seed quality and vigour, and developing a new design of steam weeder, and was connected with Tim Chamberlain's goal of kickstarting an organic carrot seed industry in Canterbury.[18]

With the trust in control, Bob directed his energies elsewhere. He had been living on his own again since late 2000 and spent a great deal of time in his half-acre garden on Ashgrove Terrace. He sometimes wondered how he had ever managed to keep the garden up *and* work full time. He started to volunteer one day a week for Geoff Barnett on his organic farm at Motukarara and enjoyed walking the property with Geoff and thinking through possibilities for development – not to mention the inevitable weeding of onions. He grew his Crowder's Delight, the globe artichokes he'd originally brought from California in 1969, and managed to keep the dry bean lines from the BHU going.[19] The arrangement worked well for both men; in 2023, at the age of 84, Bob was still visiting Geoff most Thursdays.

In return Geoff supplied Bob with fresh organic vegetables, dropping them off on Tuesdays while doing his other town deliveries. These Tuesday morning visits eventually became a regular event involving other members of the local organic community. Bob played host at Ashgrove Terrace and ensured that the coffee and cake, Windsor blue cheese, Breadman organic crackers and his own feijoa or fig chutney flowed plentifully.

•

The Green Party's success in getting into Parliament in 1999 in its own right (not, as before, as part of a broader coalition of left-wing parties) heralded an important change for the organics movement. Under the co-leadership of Jeanette Fitzsimons and Rod Donald, and with a more professional Soil & Health Association to support, the government released funds for major works as part of its confidence-and-supply arrangement with the Green Party. These included a research project on incorporating school gardening programmes into the formal curriculum, and the Small-Scale Producers Organic Programme (SSPOP) – an organic certification for small farmers. Jeanette had been involved in the drafting of BioGro standards in the early 1980s, and Rod had been an enthusiastic supporter and founding trustee of the Organic Garden City Trust in Christchurch and its school garden programme, Kids' Edible Gardens, since its inception in 1997.

With Brendan Hoare as national Soil & Health Association president and Meriel Watts as director, the association rose to the occasion. Brendan had already orchestrated the high-profile Organics 2020 event in Auckland in 2000, demonstrating that the movement was ready to take on a leadership role. Now the association received a contract for services with the government worth $300,000 – more money than it had ever seen before. Furthermore, in 2001, OFoNZ received $70,000 to develop the Organic Sector Strategy with MAF (followed, as will be seen, by over $3m in 2005 to boost the sector significantly).

The story of how these seed funds were expended is part of a much broader history of the organic movement in New Zealand, but they were a strong sign of the impact Bob's pioneering work was now beginning to have. Direct funding of this magnitude had never been seen before in the organics movement and forward momentum of this kind had not been sensed since 1988, when Bob helped to orchestrate the MAF funding for the IFOAM and NAASA delegations to New Zealand.

Bob was largely absent in this new round of support, although he did contribute to the SSPOPs. Soil & Health held the funds for this project – to establish an organic certification for non-exporting growers and manufacturers using a Participatory Guarantee System (PGS) – and the subcontract was awarded to Chris and Jenny May. Their work with Bob in the early 1980s had been critical to the establishment of BioGro, and international PGS work was by now Chris and Jenny's bread and

TOP: *OFNZ launch, 2002. Left to right: Karma Burwell (Soil & Health co-chair) Brendan Hoare, Jill Hamlin (Chair, BioGro), Terry Higginson (Chair, OFNZ), Jeanette Fitzsimons, Jim Sutton (Minister of Agriculture).* BOTTOM: *Jeanette Fitzsimons at OFNZ launch, 2002.*

butter. Bob was enthusiastic about the programme, not only because it meant small-scale organic growers could finally get recognition for their efforts without paying a relative arm and leg, but also because it opened possibilities for peer-to-peer education and support, and meant that places like community gardens and even home gardens might be able to claim a valid organic certification status. Canterbury was selected as one of four pilot areas for the programme, based on the stability of the Organic Garden City Trust as a host organisation. The scheme had 'the potential to revolutionise the way food is grown and distributed, to actually make terms like regional economies become exciting spheres of community ownership'.[20] Nationally, it was eventually launched as Organic Farm NZ (OFNZ) in November 2002, a PGS that leveraged BioGro's standards. The launch took place at Brendan's Eco Sanctuary at Unitec with Jeanette Fitzsimons and Minister of Agriculture Jim Sutton in attendance.[21]

Bob had attended more than one hui to explore what the scheme could look like and in November spoke at the launch of Canterbury Organic, the organisation that would manage Canterbury certification.[22] In 2003 he made regular trips to Golden Bay to audit a group of growers in the scheme, which included the community gardens that Sol Morgan was now running.[23] In 2004 Bob attended a small-growers' certification meeting in Auckland with Brendan.[24]

Bob did not remain involved in the national-level conversation, however. Despite the injection of funds and opportunities into the sector, from 2000 – and particularly through 2001 and 2002 – no one was really talking, publicly at least, about the BHU. From the outside it looked like the unit was still closed for business. 'That was when the BHU got forgotten,' Bob remarked. 'I was already getting a bit bitter and twisted about the lack of support for the BHU.'[25] This lack of enthusiasm seemed to Bob to be expressed from all sides: the government might be allocating small pockets of funds, but it wasn't rewriting policy; the organic movement was absorbed in new projects; and, of course, Lincoln University just seemed to want the BHU to go away. It was, he said, 'a pretty lacklustre time … It was a time of the hui rather than the do-ey.'[26]

Bob's mother Madge passed away in June 2001, eight years to the day since Bill had died. She was 93. Bob marked the occasion with a fish and chips gathering – Madge's favourite – in the sleepout at Ashgrove Terrace, a convenient place for her remaining friends to get to. 'That's what mother would have wanted. She said she didn't want any special thing.' Shortly afterwards he scattered her ashes on Roundway Hill, where he had earlier taken his father's. 'Mum's day & the sun shone brilliantly on Roundway,' he wrote. The service 'went well despite difficulty with getting [the] box of ashes open'. After a drink at the Bear Hotel, Bob and his brother David returned to Roundway: 'watched next lot of storm clouds spread down from N. *One* flash lightening back of Pewsey.'[27]

In 2004 Bob returned to Roundway where he 'cleared Mum/Dad's ashes spot of litter, walked over steep faces – plenty of orchids but only seed heads, also plenty of cowslip dead heads'.[28] Together, he and David visited their Aunt Teg in Birmingham then drove to Dover, crossed the Channel in the train and went on to Bonn.[29] Bob

The sun shone at the scattering of Madge's ashes, Roundway, 2001. Bob is centre, looking up, with David and Simon behind, to his right.

managed to spend time with his nephew Simon and his partner Sophie, including at a street party which was 'really excellent, good family fun & people to talk with', and visited his old friend and colleague, Bernward Geier.[30]

Three years later, in 2007, Bob received the news from Simon that David had died at home.[31] Bob was now the only surviving member of his immediate family.

•

Tim Jenkins was busy with harvests and experiments at the BHU and initiated a series of tours and workshops. The latter, funded by MAF's Sustainable Farming Fund, commenced in July 2002 with a session by Tim on soil testing. Future workshops were planned to cover worm farming, fertilisers, pest control, biodiversity, weed management, export marketing and a host of other topics.[32] 'It was really a huge challenge,' Tim remembered:

> We had funding but it was all geared at producing outputs from experiments and workshops etc. There was big pressure to earn all of that money and we were regularly audited for reaching the targets of the funding. I had my academic role as post doc and developing courses, and the unit itself was a forest of weeds

with a massive challenge of twitch/couch grass. We did not really have funding for renovating the unit and I think this created some confusion in the organic community. Why were we not creating some organic paradise for people to visit and be inspired by? Well, the reality was that we needed to do summer fallows (repeat cultivation over summer to deal with twitch) and put scarce resources into replacement of main line irrigation and fixing regular irrigation blowouts. We needed to create replicated experiments on such cleared and irrigated land to fulfil our funding requirements. This left little time for the cute model systems. What we did achieve though was some nice working systems that were actually commercially productive.[33]

By April 2003 the BHU had its own website.[34] In an era when an online presence meant the difference between being seen and utterly invisible, this extended the reach of the unit. Slowly but surely, the trust board was executing a plan to resurrect the unit and rebuild wider community engagement, which Tim felt was successful:

We had very large open days (sometimes hundreds of people, around 300 at the 2004 open day) and Small Farms Workshop attendance (depending on the topic and timing ranging from 20 people to over 100 people). For each workshop, a manual was written that contributed to the output of the BHU. In addition, there were monthly articles in *Canterbury Farming* and regular articles in *Organic NZ* as well as [my] regular monthly column in *NZ Gardener* ... I think this was excellent visibility, profile and highly useful information (regularly published and easily available) to support farmers in particular and gardeners also.

We also fostered organic growers, getting them established on, or inspired at, the BHU and moving on from there.[35]

In January 2004 Sir Peter Elworthy died and Lady Fiona retired from the trust board. Jon Manhire took over as chair, a position he would hold for more than 19 years.[36] In August Tim wrapped up his role at the BHU, although he remained at Lincoln in other capacities and for some years wrote articles on behalf of the BHU.[37] Tim and his wife, Vesna, even leased land at the BHU to grow organic vegetables for farmers' markets.[38] Once again, nobody was employed to oversee work at the unit.[39] A part-time general manager's role was eventually advertised in March 2005 and filled in April by Holger Kahl, who had been head of the School of Horticulture and ran the organics programme at CPIT (which later became Ara Institute of Canterbury).[40]

Bob continued to interact with the organic sector, reflecting on his gardening style and sharing knowledge and ideas. He was a regular guest speaker in Brendan Hoare's courses at Unitec in Auckland until 2005 and presented lectures at events such as at Ōamaru's Organic Food and Wine Festival, the Southern Seed Exchange's seed swaps and meetings of the Good Gardeners Association.[41] In his talk to the latter in November 2002 he explained the rationale behind his 'organic wilderness garden' of 22 years, which consisted of layers of plants from strawberries to currants to fruit trees and large nut trees: 'The rich variety of the wilderness garden aims to create balance through variety, eliminating the problems inherent in monoculture systems.'[42] In his description can be seen the permaculture basis of the food forest concept, although Bob didn't call it that. From 2006 he resurrected his writing for *New Zealand Growing Today* and *New Zealand Lifestyle Block* and began a regular gardening column for the Christchurch *Star* that continued until at least 2014.[43]

During this time Bob became more involved with the community in Golden Bay, making regular visits to his former student Sol Morgan and his friend Klaus Thoma. Klaus had worked at the Department of Agricultural Research office in South Australia and was involved in NASAA from 1983. In 1986 he and his wife Maria bought property in Golden Bay. He met Bob in the early eighties and had become one of his closest friends. On his visits, Bob offered horticultural advice as the couple developed their property. Klaus had also undertaken the original site survey for the Golden Bay Community Gardens in Tākaka.[44] In 2000 Sol Morgan became involved as a volunteer and later as a paid employee at the gardens, and regularly sought Bob's advice. Bob, he said, was 'always keen to hear what I was up to ... I always appreciated his advice.' Bob was patron of the Golden Bay Community Gardens for some time.[45]

In 2003 while on a visit to Autumn Farm, a gay accommodation venue in Tākaka, Bob met Mic Fischer, a young German Wwoofer. The next day Bob moved from Klaus's place to Autumn Farm and he and Mic spent a day pruning at a friend's place. In his diary he noted, 'Excellent day spirits raised – move into Autumn Farm – gloom lifts.' A few days later he shared 'a hot bath under stars with Mik [sic] – Nice.'[46] On a visit the following month he wrote, 'Mik very much on my mind and looking better than ever. Sigh.' Two days later the pair shared another outdoor bath, and the following night Mic joined Bob in his bed: 'Bliss but not the best sleep.'[47] A phone call from Mic eight months later brought up 'emotional feelings'.[48]

Klaus's property became certified organic through the OFNZ scheme in 2005, and Bob became more involved – especially once Klaus and Maria shifted the focus of their production to vegetables in 2007. His travel diaries record regular trips to Golden Bay between 2003 and 2008, sometimes every few weeks, and he assisted with direct financial support as well as physical labour and practical advice. Klaus and Bob shared a deep philosophical connection and together lamented the 'neo-liberal features of organics'. Klaus considered Bob 'a close friend ... and sometimes ... even a little bit of a father figure ... His whole career was basically speculated … on this thing [New Zealand as an organic nation] coming to life … He unfortunately carried a can on getting the organics world established.'[49] Bob's connection with the community gardens, the organic shop in Tākaka and the gay men at Autumn Farm spoke to values of localisation, community and diversity that were far stronger measures of progress than economic output.

Bob's trips to Golden Bay dwindled in 2008 after an argument with Klaus, though he continued to visit and help on the property at least a couple of times a year. He spent his birthday in Tākaka with Klaus and Maria in 2012 and wrote, 'I do respect Klaus as a good friend. I have known him for so long now.'[50]

Within the organic movement, Bob continued to support his favourites such as Brendan Hoare, who joined the IFOAM World Board in 2005 at the Adelaide meeting.[51] Brendan recalled, 'When I first got onto IFOAM he gave me a koha … [He was always] very generous, loving and generous, and wanting you to succeed … Bob has demonstrated the type of leadership that I've always really enjoyed.'[52]

At the IFOAM meeting in Adelaide, Brendan and Steffan Browning (later a Green Party MP) lobbied Green MP Sue Kedgley for broader support of the organic movement in New Zealand. As a result, the government funded the establishment of Organics Aotearoa New Zealand (OANZ) to the tune of more than $1.5m.[53]

The Organic Advisory Programme (OAP), a free advisory service to support conventional farmers to transition to organics, was launched in 2006 with a budget of $2.2m. Jon Manhire saw that the BHU 'would be in a strong position to deliver such an advisory service and could obtain significant funding to do so'.[54] He and Holger Kahl were involved in the design and subsequent implementation of the OAP, and Holger was appointed South Island regional advisor in 2007.[55] Holger saw 'strong synergies for the BHU and the OAP, for example in holding workshops and running demonstrations', and the South Island OAP essentially ran out of the unit.[56] The OAP lapsed in 2009, but in that time evaluations of the programme were positive, even if, as Holger felt, 'the three year timeframe was way too short'.[57] Overall, 150 on-farm consultations were conducted nationally; two thirds of these farms went on to convert to organics.[58]

Holger had resigned from his part-time role at the BHU, and with CPIT's organic programme on hold he focused on developing the concept of a BHU-run Organic College, first flagged the year before.[59] In an example of how quickly the organic movement was maturing, he was successful in securing $30,000 from OANZ to co-fund the college with Work and Income New Zealand. Lincoln University's vice-chancellor initially poured cold water on the concept, but a twist of fate, which saw Telford Rural Polytechnic (which had supported the initial proposal) become part of Lincoln University, made the establishment of the college possible.[60] The

BHU Organic Training College was successfully implemented in 2007 and created an entry point for Bob back into the BHU.

On 19 July 2008 Bob travelled to Wellington for BioGro's 25th anniversary. He found the afternoon AGM 'uneventful', and refreshments were served in the BioGro offices before a photo session. Bob spoke with Tom Lambie, an organic farmer who was now chancellor of Lincoln University: 'amazingly free & easy with his comments, thoughts and aspirations'.[61] The evening dinner event was formal. Bob was seated at the top table and 'treated almost as [a] celebrity'. Speakers included CEO of Organic Aotearoa New Zealand Jon Tanner and James Millton of Millton Vineyards, with whom Bob had a 'good chat'. A 'small presentation' was made to Bob in honour of his services: 'I replied briefly.' If there was any damper on the evening for Bob it was that Chris and Jenny May were 'not treated quite so well as Perry & I – felt a bit embarrassed really as those two were such an integral part of it all'. He walked back to his hotel with James Millton and Jared White (now a BioGro auditor) around midnight. Jared went off with some others for a drink: 'Us oldies went to bed.'[62] At 69, Bob no longer felt like partying into the small hours.

In November that year Bob attended the BioGro 25th anniversary Southland regional dinner, held in Gore, where he spoke to the gathering of 40–50 people.[63] The next day he drove to Riverton where he visited Robert Guyton and his son Adam at the Environment Centre before joining them later for dinner. The Guytons' property was one of the first food forests to have been established in New Zealand. On a walk around Bob observed, 'impressive growth – runner beans in flower – all very advanced even compared with Christchurch.' Bob visited again the following day to take photos and returned to the environment centre to meet Robert's partner, Robyn.[64]

By this time Bob's relationship with the BHU had warmed a little. Bill Martin, another former student, was appointed manager of the unit in late 2007.[65] Bill's job was not only to take care of the BHU but also to establish the Organic Training College there. He was well placed to do this work. His studies had started at Massey University with diplomas in horticulture and small business (1987 and 1989), and in 1990 he completed a Certificate in Ecological Horticulture at the University of California, Santa Cruz.[66] He worked as an intern at the BHU with Bob and Geoff during the summer of 1993/94 while studying for his Bachelor of Science in ecology at Lincoln, which he completed in 1995. In 1994 he was appointed the organics technician at the Seven Oaks campus of what was then Christchurch Polytechnic, working alongside Holger Kahl. In 2007, Bill became head of the School of Horticulture, but could see the writing on the wall.[67] Seven Oaks closed down in 2008, and Bill was allowed to bring across many of the teaching resources and teaching staff from Seven Oaks to the BHU. It looked as if there could be a revival.

Bob became involved in developing the teaching programme at the BHU Organic Training College and for several years he had a run of six or seven lectures in the programme. His teaching work continued until 2015, when he and Bill had a falling out. Thereafter, Bob gave only occasional guest lectures, usually on the history of organics in New Zealand and the BHU, and did a walk-around with the students in which he challenged them to think critically about what they could see.[68]

In 2010 Bob was invited to become patron of the Soil & Health Association of New Zealand, a role he shared with Jeanette Fitzsimons, whom he held in high esteem. Bob had first met Jeanette in the early days of BioGro when the standards were being developed.[69] Since then she had served as the Green Party co-leader from 1995 to 2009 and a Member of Parliament from 1996 to 2010. On her retirement from politics, the Soil & Health Association saw a significant opportunity. The national council was keen to utilise its patrons, something that hadn't been done since the time of Sir Dove-Meyer Robinson in the 1980s. Inviting these two powerhouses of the organics movement to take up these roles was a strategic move to lift the profile of an organisation that had been drifting somewhat since the excitement of the early 2000s.

During the time Bob was patron, the national council was focused on fighting fires in court around genetic engineering; introducing and formalising biculturalism and a celebration of te ao Māori worldviews within the organisation; and, last but not least, attending to repeated overtures from BioGro to be allowed to come under the umbrella of Soil & Health and dissolve its own parent body, the New Zealand Biological Producers and Consumers Council (NZBPCC, formerly the NZBPC), which Bob had worked so hard to help establish in its early years.

Having Bob and Jeanette's names on the national appeals to support the court cases around genetic engineering was critical to the organisation's various successes in this arena. The cases, essentially fighting appeals from Federated Farmers (the earlier collegiality had long since evaporated), required ongoing funding, and the appeals raised tens of thousands of dollars. Genetically engineered crops on New Zealand soil were held at bay a little longer, something Soil & Health considered an essential precondition to achieving its vision of an organic Aotearoa New Zealand.

As patron, Bob was supportive as ever but did not drive decision-making. Privately, he believed that the growing of GE crops on New Zealand soils was inevitable, as science cannot be stopped. He was characteristically pragmatic about BioGro: he considered its proposition to join Soil & Health on its present merits, rather than on historical recollections that might interfere with a rational decision.

BioGro's prompting in fact helped Soil & Health realise that it no longer had a clear strategy, and the development of this was prioritised ahead of other decisions. It wasn't until after Bob had stepped down as patron that the NZBPCC dissolved itself, and its company, BioGro, was handed to Soil & Health (ostensibly as an asset).

Bob enjoyed being able to take on a kind of 'governor-general' role within the movement: not directing policy or sitting in board meetings, but debating, guiding and supporting the movement. He was deeply respectful of Jeanette Fitzsimons' practical approach, something she described herself at a public meeting to welcome her to the role: 'I don't come to it with starry-eyed idealism, having done my time repeatedly over 30 years on the end of a gorse grubber, looked for solutions to barber pole worms in our lambs and revelled in being able to grow almost all our own food.'[70] Around 80 people came to Jeanette's talk, and Bob noted that Margaret Jones, a previous patron and well-loved within the organics fraternity, 'at near 90 [was] a good hit'. They went to a pub afterwards that was 'full of young people, noisy but very friendly. Margaret a great hit there as well.'[71]

Records show that Bob and Jeanette were actively involved in national council meetings and took turns to attend, depending on where the meeting was held. Bob travelled to Nelson in August 2011 and participated in what he called a 'good debate'.[72] In 2014 he was confirmed in the role of patron for a further two years.[73] Remits to the 2016 Soil & Health AGM removed the requirement for the association to always have a patron; to date, Bob Crowder was the last.

•

The first of the earthquakes that shook the Canterbury region for over two years woke residents in the early hours of 4 September 2010 and caused widespread power outages, disruption to water mains and sewage services and significant damage to the central city. In the confusion of trying to find a torch in the dark, Bob bumped and cut his head and stood on a fallen photograph, shattering the glass. He was in shock and over the next few days lost several kilos of weight. But in a way, the earthquakes helped him to reframe his purpose and his understanding of the movement he had helped to re-establish, and of which he was now patron:

> All the evidence, both anecdotal and scientific, indicates that a truly organic environment has the ability to buffer the system against adversity and deliver the goods in time of need. Unfortunately, such a system still does not sit comfortably within the existing economic paradigm, but times they are a-changing so rapidly that an alternative economic paradigm will just have to come about and sooner rather than later.[74]

He reflected with satisfaction that he had plenty of stored water and preserves, and his own potatoes, onions, garlic and dried beans in the cupboard. His garden was full of carrots, parsnips and leeks, and 'rich with the spring flush of salads and even the first asparagus thrusting up into the spring sunshine'. The event confirmed for him 'the need to expand the concept of sustainability and self-sufficiency into our total way of life … Hopefully our stress can be the nation's wake-up call and holistically organic by 2020 can still come to pass'.[75]

The next major earthquake struck in February 2011 and caused 181 deaths and building failure throughout the central city. Eighty-five percent of structures in the CBD were subsequently demolished, and 8000 homes were cleared in a 'residential red zone' that stretched from the central city to the ocean along the corridor of Ōtākaro Avon River. This time the effects on Bob were marked. 'I think it activated all my post-traumatic stress syndrome … [my reaction] may well have been a result of being bombed during the Second World War … it reawakened those kinds of fears in me.'[76] He also blamed the stress for another bout of the 'Tapanui Flu' he'd contracted in 1982. This illness, which is now called myalgic encephalomyelitis (ME) or chronic fatigue syndrome, can be debilitating. Bob headed out of town to stay with hazelnut farmers Jim and Eleanor Jolly in Geraldine. He worked with them for a week and returned a few days later to help lay nets for the nut harvest.[77] In his words, the earthquake 'did quieten me down a bit'.[78]

The Christchurch central city was surrounded by razor wire and a military cordon for months, and Bob was keen to get away. He returned to Britain for the British summer of 2011 and had the loan of a house in Devizes. He repeated the journey in 2012 and 2013 and spoke of this as 'a very inspiring type of period … It was a very social time'.[79] His diaries describe happy days in and around Devizes, visiting old friends and walking familiar paths: up One Tree Hill to see the bluebells, wood anemones, celandines and primroses, and Roundway Hill for cowslips and early orchids and to pay a 'silent tribute to Mum & Dad'.[80]

He encountered a few disappointments while there: Martin Stokes emailed Bob to say he wasn't interested in meeting, and Steven Craik failed to respond to messages. Even Mark Measures at Elm Farm seemed reluctant to talk to him at first. But these rejections didn't dampen Bob's spirits too much. He managed a visit with Chris and Allan, gave a talk on the BHU at Elm Farm, and visited his old student John Calvert at his Butterfly Farm in Stratford-upon-Avon.[81] He found time for family too: he flew from Britain to Kohn in 2013 to stay with Simon and Sophie and their son Hugo. 'Hugo greeted me well but finds it all very strange and suspicious.' Bob looked after Hugo one morning on his own and managed perfectly well – during a shopping trip he found 'Hugo excellent … really quite easy to manage'.[82]

While in Kohn he visited the IFOAM office. Bob had left the World Board in 1998 but retained an active interest. On this visit he attended a staff meeting and met the new executive director, Marcus: 'warmed to him, he appears an excellent person for the job'. Bob was asked to give a brief overview of the organic scene in New Zealand. To cap off a wonderful day, he had beer and chips in a nearby beer garden with Simon, Sophie and Hugo.[83]

The earthquakes created an entirely new reality for the people of Canterbury, especially for residents of Christchurch, and left a large number of people physically injured and a looming mental health crisis. This was not the environmental disaster Bob had long warned against, but the impacts were similar and the answers the same. Suddenly, there was urgency around developing resilient urban systems.

There was also the immediate question of what to do with land now deemed unsuitable for habitation. The Christchurch 'red zone', an enormous stretch of land along the river corridor, fell into this category. The Canterbury branch of Soil & Health called a meeting to discuss possibilities in 2012. Around 40 people attended, and the meeting expressed a strong desire to make use of this land for growing food. If Christchurch was to become resilient to crises – and it needed to – food had to be grown in the city on a large scale.

A follow-up workshop was held on the spring equinox of 2013 at the University of Canterbury, to expand this vision and bring relevant agencies around the table who could make it happen. Among those present were staff from the Christchurch City Council and Waimakariri District Council, coordinators of community gardens, members of the new Christchurch Food Forest Collective, the Tree Crops Association and more. The event was co-hosted by Soil & Health Canterbury and the Rangiora Earthquake Express.

Bob spoke passionately at this workshop. He recommended that any region-wide initiative to grow food in public places should utilise the BHU as a place where people could see, and begin to understand, what a sustainable food system looked like. He recounted how the BHU was inspired by horticultural students who wanted 'something sustainable and organic taught at Lincoln University'; it had been 'devised as a total rebuilding of the horticulture movement'. He told of the way the community had donated time and plants to establish the unit, and how '[t]housands of dollars from volunteer subscriptions and business donations went into it – primarily through Soil & Health's Project GRO fund in the 1980s'.[84] The workshop agreed to the immediate need to establish a steering group to realise a vision for a patchwork of food forests woven into the community, and the BHU was acknowledged as one in a network of vital 'nodes' to make this a reality.

The first meeting of the nascent steering group took place at the BHU two months later. Jon Manhire welcomed participants, BHU students fed them, and Bob took them on a tour of the unit. The minutes from this meeting speak to the enthusiasm for establishing an entirely new possibility for the earthquake-ravaged city, building on ideas that Bob had propounded for decades:

> Bob led us round the BHU, which was a huge honour. He emphasised how it was imperative ... to take a holistic approach to landscape rebuilding. He illustrated how social values had influenced the landscape and projects at the BHU over 30 plus years. As we walked around the BHU Bob pointed out a large diversity of projects ... He illustrated just how important it is to have a long-term vision when redesigning landscape, so that in decades to come the initial vision can be related to future generations. Society, psychology and biology are intertwined ... [We] have an opportunity to sow seeds of different ways to achieve food production.[85]

A follow-up workshop was held at the University of Canterbury on 22 March 2014. The movement – for that is what it was becoming – was already referred to as the Food Resilience Network, a name that was formalised in June.[86] A large amount of work emerged quickly from this movement, including new policy for the Christchurch City Council and an opportunity to create a high-profile food resilience centre in the central city.

Bob chose not to attend any further meetings after July, saying, 'better to just be an observer, I think'.[87] He retained an interest, however, and continued to offer support and advice. His August 2014 email to Tony Moore, Sustainability Advisor for the Christchurch City Council, who was pivotal to the nascent movement, provides a snapshot of the keen excitement he felt for the project. Bob imagined food trees merged into existing tree plantings, 'starting with the approach to Botanic Gardens car park and then onto and around Hagley Park'. The cherry trees of Harper Avenue could be underplanted with violets, celandine and miner's lettuce. 'If this can all be related to the Botanic Gardens then you have the opportunity to create an educational experience.' He explained what he knew of disease-resistant apple varieties, as researched at the BHU.[88]

Bob saw an opportunity once again to bring his practical work at the BHU to the wider community. His connection with a revitalised local food resilience or food sovereignty movement in post-earthquake Christchurch began to centre on support for younger people working in that space. One of these, Bailey Peryman, initiated a range of practical urban farming projects starting with Agropolis – a community garden using raised beds on a site of compacted rubble on High Street in the central city. The optics of this project – a resilient and sustainable garden city emerging from the earthquake rubble – were powerful. Bailey later co-initiated Cultivate Christchurch, a social enterprise providing work experience and educational opportunities to young people not in education, employment or training (YNEETs). He also was one of the key movers of the emergent Food Resilience Network and its flagship project, Ōtākaro Orchard.

Projects like Agropolis, Cultivate Christchurch and Ōtākaro Orchard attracted others to the city. Peter Wells arrived in early 2016; his connection was through the Beacon Food Forest in Seattle. A mutual link through anthropologist David Border Giles, who spoke at the 2013 foundational Food Resilience Network hui, ensured that Peter made contact with Bob, with whom he stayed for a few days. Peter recalled, 'between Bob's stories and deep history of the national organic movement ... the electricity of the city pulled me in'. Eight months later he returned and would stay for five years, working with the Food Resilience Network, 'stewarding the Ōtākaro Orchard project, and a governance term with the Soil & Health Association'. Peter recalled Tuesday morning coffees at Bob's: 'There was always a robust discussion of the latest developments in the farming sector and Bob *always* had observations to report about the fluctuations of the garden and fervent notation of the rain gauge after every event.'[89]

•

The BHU continued with Jon Manhire as chair, Bill Martin managing the Organic Training College and Charles Merfield running the Future Farming Centre. Charles had initially been employed for two days a week to develop the research programme at the BHU, which had expanded into a formalised Future Farming Centre under his leadership in October 2011.[90] Shortly after Charles started in this role, Bob tasked him with resolving the psyllid problem for the organic growers. Charles had already worked with mesh crop covers in the UK and trialled this. It worked: he found he could control the psyllid almost completely, better than any agrichemicals.[91] The psyllid work went on for seven years through to 2018.

The possibility of branding the work of the BHU had been broached in 2005, when the board agreed that 'many people do not understand what the names BHU or Biological Husbandry Unit stand for'.[92] The same meeting agreed to a new mission statement for the BHU: 'to research, demonstrate and promote organic systems through education and training, research and development, demonstration models and consultancy'.[93] Ten years later BHU staff were involved in hui with hapū around Banks Peninsula that focused on growing food at Koukourārata Port Levy, and specifically on growing taewa ('Māori' potatoes). Vision Mātauranga: Koukourārata, which was deeply consultative, led to sufficient funding to employ Charles Merfield to help develop the growing programme full time for two years.

Although the initial project was successful, a follow-up project failed and, with funding spent or returned, Charles left in 2019.[94] However, in 2020 a contract from the Ministry of Education to provide BHU biodiversity kits translated into te reo Māori changed all this. The success of the project enabled Charles to return in April 2020. He helped to assemble the kits and execute the project, and then caught up on maintenance work on the farm itself.[95] From 2021 there was something of a revitalisation of the Future Farming Centre; after 10 years of building the brand new contracts were beginning to come through.[96]

Jon Manhire acknowledged that funding for the BHU remained a problem; it had been hard to 'sustain baseline capital to maintain infrastructure'. However, for now, the BHU was on 'the right side of the ledger'. A review document for the BHU had been written, which, Jon felt, contained an exciting and aspirational plan. Bob's vision for housing on site – an evolution of the eco-village idea – was on the cards at long last, albeit it at the bottom of a long list of priorities.[97] The original lease, signed in 2002, was about to expire, but Lincoln University's vice-chancellor intended to renew it.[98] A 15-year extension was, in fact, confirmed in November 2021.[99]

•

With his energy, credibility and resources, Bob Crowder made an immense contribution to the development of New Zealand's organic sector. In 2021 that sector was valued in the hundreds of millions, as the 2021 Organics Aotearoa New Zealand (OANZ) report showed:

> The organic sector generates approximately $620 million in export and domestic market revenue, with a further $100 million worth of products imported into New Zealand to meet consumer demand. This represents an average annual growth rate for organics of 6.4% for the past three years, without incentives or supporting policy frameworks. With just 86,000ha under organic certification, this amounts to average earnings of about $7250 per hectare – responding to consumer-driven demand for organic products, particularly organic dairy, wine and kiwifruit. The sector exports to many global markets, including those in emerging economies where a rising middle class is using their new economic power to buy better food for their families.[100]

The numbers are impressive, but more impressive still is the fact that an organisation such as OANZ exists at all to commission this report. It is a direct outcome of the effort Bob put into developing the sector, especially in the 1980s. He was not alone in this, of course, but those who were building the sector at that time agree that Bob's part was critical.

10: RETIREMENT, 2001–23

Yet, in 2021, despite the state of the BHU, the increasing success of organics in New Zealand and the growing financial value of the organic sector, Bob was disheartened. Nearly 30 years had passed since the United Nations Conference on Environment and Development, the 'Earth Summit' of 1992. Warnings about catastrophic climate change had been ignored and the world was in peril. Things were worse, not better.

He regretted not trying to do more in his retirement: 'Looking back, I could have used my time a lot better ... I often think of [my retirement] and think I haven't made the most of it. But ... it's a result of total frustration with the 30 or 40 years prior to it of constantly banging your head against brick walls, and sleeping on people's floors.' There was that double wrong: the work had been thankless *and* fruitless. The oily rag that had powered the organic movement was not recognised. He recalled Bernward Geier, for so long the director of IFOAM: 'His house was constantly filled with people because everybody was in the same boat ... we all would sleep makeshift in various parts of people's houses.' His own work, he felt, had lost momentum after the IFOAM conference. 'After IFOAM '94 I just thought to hell with it, I'm just going to relax and do nothing.'[101] His mother's well-intentioned words of 1983 were still apt: speaking with reference to Bob's organic exhibition at the A&P show, she said, 'Don't get downhearted if it doesn't go as well as you hope, it takes a long time to alter people's way of thinking and living.'[102] But he *was* downhearted, and his views reflect his own high standards and a painful sense of failure. But Bob had certainly not 'relaxed and done nothing' since his retirement.

Bob's list of achievements is monumental: he helped to ignite a movement that continues to inspire and capture the imagination of people of all ages in Aotearoa New Zealand. And he did it in the face of intense opposition and prejudice. Charles Merfield commented on this point:

> It's hard to conceive of what Bob was up against in the 1970s when he converted to organics and started the BHU ... The 1970s was the height of intensive agriculture. The solution to everything was pesticides and fertilisers, and so having the strength of character to swim against an incredibly strong current in agriculture ... really speaks ... massive testimonies ... For Bob to have gone against that two decades prior [to more public understanding in the 1990s] really is quite astonishing and testament to his determination to keep going despite the negativity – especially the negativity from other Lincoln academics ... There were people who still refused to shake his hand after his Queen's gong, and after he was retiring, there were people who would still refuse to engage with him.[103]

Bob's legacy is not often found in the published records of formal institutions, which he generally eschewed. Truth be told, he had an aversion to academia and the perceived hypocrisy of people who wrote one thing but did another. Bob has been a man of action, and his legacy resides in the hearts and minds of those many, many people with whom he so generously shared his life; it rests in the glimpse he created at the BHU of a world beyond the neo-liberal economic paradigm. As he declared in 1999:

> We've come a long way in organics since I started in it. What's lacking today however is the true spirit of organics, the holistic relationship with the living earth. I want you to remember to practise what you preach. Don't lose the philosophy behind organics.[104]

There may be irony in the fact that his legacy can also be seen in the dollar value of the huge export market he helped develop. This has been referred to as the 'organic dilemma' – the contradiction of a developing organic market for wealthy Westerners that tends to reinforce a paradigm that perpetuates injustice.[105] A booming organic market does not necessarily meet the aspirations of those who simply want to see all people with a good supply of health-giving foods and a natural environment where living things thrive. It is a tension that has long sat at the heart of the organic movement, and Bob has not been oblivious to it. He has actively wrestled with it, at times pronouncing 'It's no good: we're all doomed!' while continuing to host gatherings of kindred spirits who keep making the world better because it's the right thing to do.

In 2021 Bob was still firm about his vision for the BHU:

> Organics is nothing without an understanding of total ecology and environmental biodiversity. Biodiversity is the key word now and they're starting to realise this, belatedly. And so, it's not a question of setting up a commercial unit in the idiom and paradigm of today. It's visualising what our society has got to be if there is going to be a meaningful change in how we treat the total environment. Until that happens, the BHU will struggle on like it is at the moment, with no real purpose in mind.

> By now, that place should be an eco-village. It was never meant to be a research station. It was meant to show how ... biodiversity was necessary in order to be able to have an organic system. People don't understand that, never have done. Certainly, Lincoln doesn't.[106]

His vision reached beyond the BHU, of course. New Zealand needed to become 'a certified organic nation', a move that 'would solve most of the environmental problems'. The organic standards ensured environmental integration, biodiversity enhancement and improvements in the lives of human beings.[107]

In 2022 he explained this to students at the Organic Training College:

My concept of organics is not just growing vegetables with compost, it is … changing the whole philosophy of how we look at the land, and the environment, and everything that goes with it, and somehow fit a new paradigm of economy into it, and that is even more anathema to what is going on at a place like this [Lincoln University]. And that is why they still don't really like the idea of working within the framework of environmental systems. And so, rather than … thinking of oneself as [doing] organic growing, I like to think that we would think of ourselves as trying to have a holistic approach where we integrate everything into how we manage the land.[108]

At the end of 2022 Bob sold his home in Ashgrove Terrace to the son of his old friend Ernst Frei. At 83 years old, he was plagued by arthritis, and the back problems that had troubled him since childhood were ever present. He once quipped that the ideal age to die was the Biblical 'threescore years and ten', but he was well past that now. And his preferred cause of death, naturally enough, was to be struck by a bolt of lightning. From his new apartment on the hill he was able to look out over his old garden and the city, content now to watch the weather as it rolled in from all directions and to gaze across the Canterbury Plains to the Southern Alps where he had spent so many precious hours in his younger years.

In 2023, the government passed the Organic Products and Production Bill, creating a legal status for organics and ensuring a minimum national organic standard; and Lincoln University appointed a full-time sustainability officer. Bob thought that perhaps his work had paid off after all. However, the BHU management team announced they had decided to shift their activities into the Residential Red Zone, to join forces from 2024 with the new Climate Action Campus that Vicki Buck had initiated. It was indeed the end of an era.

Bob Crowder's influence in New Zealand's organic movement is marked. While some saw him as abrasive and dogmatic, others were deeply touched by his sincerity, enthusiasm, sensitivity, loyalty, generosity and kindness. His vision for and demonstration of a balanced and vibrant world measured in happiness and health for all was inspirational and, for many, life-changing. He remained firmly committed to the role of education in changing the world. But it was education of a particular type. He followed Plutarch's maxim: 'The mind is not a vessel to be filled but a fire to be kindled.'[109]

Brendan Hoare articulated what was a common sentiment. Coming to the BHU was, he said, like 'moving into … a studio, and the leader of the studio enabled you to be who you were … And the studio was called the BHU, and the lead artist mentor was Bob Crowder … He allowed my spirit to flourish, and I'm forever grateful for that.'[110]

This may be Bob Crowder's most enduring legacy.

Notes

PREFACE

1. There has been some debate over whether New Zealand or Australia was home to the world's first organic farming society. John Paul states that the New Zealand claim is dubious, because its society did not include the word 'organic' in its name. His pick is the Australian Organic Farming and Gardening Society (1944). John Paul, 'The lost history of organic farming in Australia', *Journal of Organic Systems*, vol. 3, no. 2, 2008, pp. 2–17.
2. Bob later learned that the scandal had a direct connection with his school, but said that it did not directly impact him. Bob Crowder, pers. comm., 28 July 2022.
3. See, for example, Isaac Sohn Leslie, 'Queer farmers: Sexuality and the transition to sustainable agriculture', *Rural Sociology*, 82, no. 4, 2017, pp. 747–71.
4. Bob Crowder, quoted in Scott McVarish, *The Greening of New Zealand: New Zealanders' visions of green alternatives* (Auckland: Random Century, 1992), p. 49.
5. Ibid., p. 48.

1: BRITAIN, 1939–62

1. 'Monthly Weather Report of the Meteorological Office', vol. 56, no. 1 (London, 1939): www.metoffice.gov.uk/binaries/content/assets/mohippo/pdf/a/jan1939.pdf
2. Bob Crowder, interview, 18 September 2018.
3. Bob Crowder, interview, 25 September 2018.
4. His mother gave varying accounts of this.
5. Bob Crowder, interview, 18 September 2018.
6. US troops first arrived in Britain in January 1942.
7. Bob Crowder, 'My brother: A reflection on our lives together and apart' (n.d.); Bob Crowder, interview, 18 September 2018.
8. Madge Crowder to Emily Boulcott, 10 April 1945.
9. A postcard to Emily Boulcott from Madge survives, which appears to be dated October 1949 (the date stamp is not clear). The postcard was addressed to their house in Medina Avenue, Shide, and sent from Solihull where presumably they were visiting the Crowders. This may indicate that in fact the family shifted to Devizes in 1950, but it is impossible to be sure. Madge Crowder to Emily Boulcott, 23 October 1949.
10. Bob Crowder, interview, 18 October 2021.
11. Madge recalled their wartime trips to Gurnard: Madge Crowder to Bob Crowder, 18 July 1977.
12. Bob Crowder, interview, 1 October 2018.
13. Ordnance Survey 1:25,000 maps of Great Britain, 1937–1961: https://maps.nls.uk/views/95752792
14. Bob Crowder, interview, 25 September 2018.
15. www.winter1947.co.uk/Pages/Monthly%20Summaries/February%201947%20Summary.htm

16. Cedric Roberts, 'A Halesowen Winter': www.winter1947.co.uk/Pages/Halesowen/Halesowen%20February1947.htm
17. Bob Crowder, interview, 1 October 2018.
18. 'The Weather of March 1947', *Weather Magazine vol. II* (1947): www.winter1947.co.uk/Pages/Monthly%20Summaries/March%201947%20Summary.htm
19. 'The Weather of 1947 in Great Britain', rmets.onlinelibrary.wiley.com/doi/pdf/10.1002/j.1477-8696.1948.tb00856.x
20. Bob Crowder, interview, 10 September 2018.
21. Bob Crowder, interview, 27 March 2023.
22. Lalange Snow, *War Gardens: A journey through conflict in search of calm* (London: Quercus, 2018).
23. Mikhail, quoted in Snow, *War Gardens*, p. 122.
24. Snow, *War Gardens*, p. 355.
25. Bob Crowder, interview, 10 September 2018.
26. Bob Crowder, interview, 1 October 2018.
27. Bob Crowder, interview, 10 September 2018.
28. A reference to this house can be found at: www.rightmove.co.uk/house-prices/detailMatching.html?prop=28233259&sale=66703995&country=england
29. Bob Crowder, interview, 1 October 2018.
30. Bob Crowder, interview, 25 March 2019.
31. 'Wiltshire Community History: Devizes Census Information' (Wiltshire Council): https://history.wiltshire.gov.uk/community/getcensus.php?id=89
32. 'Devizes', Britannica: www.britannica.com/place/Devizes. Wadsworth Brewery is a classic example. Bob recalled beer being delivered around the town by horse, and collecting the horse manure when he could. Bob Crowder, pers. comm., 28 July 2022.
33. Bob Crowder, interview, 25 September 2018.
34. He started attending the Patricia Scott School of Dance. He won medals in ballroom dancing and took over tuition of the ballroom dancing club at school. Bob Crowder, pers. comm., 28 July 2022. See also Bob Crowder, 'Dancing and meteorology', in Lorna Haycock (ed.), *On the Crest of the Hill: Devizes Grammar School, 1906–1969*, (Salisbury: Hobnob Press, 2006), p. 83.
35. Bob Crowder, interview, 20 November 2018.
36. Derek Jarman, *At Your Own Risk: A saint's testament* (London: Vintage, 1992), p. 22.
37. Ibid., p. 25.
38. Ironically, perhaps, the Montagu case set the stage for homosexual law reform in Britain. Neil Miller, *Out of the Past: Gay and lesbian history from 1869 to the present* (London: Vintage, 1995), p. 283.
39. David Pickering to Bob Crowder, 25 September 2011.
40. Sigmund Freud, '"Civilized" sexual morality', *Civilization, Society and Religion*, vol. 12 (London: Penguin, 1985), p. 39.
41. Jonathan Dollimore, *Sexual Dissidence: Augustine to Wilde, Freud to Foucault* (New York: Oxford University Press, 1991), p. 105.
42. Freud, '"Civilized" sexual morality', pp. 43, 45.
43. A Stevenson screen is a shelter or an enclosure to protect meteorological instruments against precipitation and direct heat radiation from outside sources, while still allowing air to circulate freely around them.

44. Robert Crowder, Diary, 1 January 1956 (Bob Crowder collection).
45. Robert Crowder, Diary, 4 January 1956 (Bob Crowder collection).
46. Robert Crowder, Diary, 20, 23 January 1956 (Bob Crowder collection).
47. Robert Crowder, Diary, 29 January 1956 (Bob Crowder collection).
48. Robert A. Crowder, 'March Weather', *The Wiltshire Gazette*, 12 April 1956, p. 6.
49. Ibid.
50. Bob Crowder, interview, 9 October 2018.
51. Bob Crowder, interview, 1 October 2018.
52. Ibid.
53. Bob Crowder, interview, 20 November 2018.
54. Bob Crowder, interview, 9 October 2018.
55. Bob Crowder, 'Dancing and meteorology', in Haycock (ed.), *On the Crest of the Hill*, p. 83.
56. Robert Crowder, 'Shivering April', *The Wiltshire Gazette*, 8 May 1958.
57. Graham Hancock, Diary, 1958. Graham wrote this shortly after the trip and sent the pages to Bob in 2011. Graham Hancock to Bob Crowder, 21 July 2011.
58. Robert Crowder to Mr & Mrs S.J. Crowder, 30 July 1958; 27 July 1958.
59. Bob Crowder, interview, 25 March 2019.
60. Ibid.
61. See, for example, Robert Crowder to Mr & Mrs S.J. Crowder, 19 June 1959, and Bob Crowder to Mr & Mrs S.J. Crowder, 19 April 1960. Bob Crowder Collection.
62. Bob Crowder, pers. comm., 28 July 2022.
63. Bob Crowder to Mrs S.J. Crowder, 15 July 1962.
64. Bob Crowder to Mrs S.J. Crowder, July 1962 (date stamp partly cut off).
65. Bob Crowder to Major & Mrs Williamson, 23 July 1962.
66. Bob Crowder to Jack Farren, 23 July 1962.
67. Bob Crowder to Mr & Mrs S.J. Crowder, 24 July 1962.
68. Bob Crowder, interview, 20 November 2018.

2: AUCKLAND AND PUKEKOHE, 1962–66

1. Bob Crowder, Book of Weather, 11 November 1961–17 March 1962, 4 December 1962.
2. Ibid.
3. Madge Crowder to Bob Crowder, 5 December 1962.
4. Bill Crowder to Bob Crowder, 6 December 1962.
5. Bob Crowder, Book of Weather, 11 November 1961–17 March 1962, 4 December 1962.
6. Ibid.
7. Ibid., 7 December 1962.
8. Ibid., 9 December 1962.
9. Ibid., 10 December 1962.
10. Ibid., 11 December 1962.
11. Ibid., 12 December 1962.
12. Ibid., 13 December 1962.
13. Ibid., 14 December 1962.
14. Ibid., 12 December 1962.

15. Ibid., 15 December 1962.
16. Bob Crowder, Book of Weather, 11 November 1961–17 March 1962, 17 December 1962. This passage surprised Bob in later years, particularly since he never smoked and would not have had a light. He felt the words in the diary were 'a disturbed mind recording the event'. Bob Crowder, pers. comm., 28 July 2022.
17. Ibid., 21 December 1962.
18. Ibid., 30 December 1962.
19. Ibid., 30 December 1962.
20. Ibid., 3 January 1963.
21. Ibid., 7 January 1963.
22. Ibid., 8 January 1963.
23. David Dennis, interview, 23 April 2019.
24. Bob Crowder, Book of Weather, 11 November 1961–17 March 1962, 8 January 1963.
25. Madge Crowder to Bob Crowder, 27 January 1963.
26. Bill Crowder to Bob Crowder, 26 January 1963.
27. Bob Crowder, Book of Weather, 11 November 1961–17 March 1962, 12 January 1963.
28. Ibid., 8 January 1963. (Here the diary reverts back to a proper weather journal).
29. Bill Crowder to Bob Crowder, 26 January 1963.
30. David Dennis, interview, 23 April 2019. The psychoanalyst was Eva Fischmann.
31. Bill Crowder to Bob Crowder, 26 January 1963; Madge Crowder to Bob Crowder, 19 March 1963.
32. Bill Crowder to Bob Crowder, 8 March 1963; Madge Crowder to Bob Crowder, 19 March 1963.
33. Madge Crowder to Bob Crowder, 8 May 1963.
34. Madge Crowder to Bob Crowder, 3 April 1963.
35. Madge Crowder to Bob Crowder, 19, 20 March 1963.
36. Madge referred to the news that Bob had shifted to another flat. Madge Crowder to Bob Crowder, 26 June 1963.
37. Bob Crowder, interview, 27 August 2018.
38. David Dennis, interview, 23 April 2019; Bob Crowder, interview, 27 August 2018.
39. David Dennis, interview, 23 April 2019.
40. Bob Crowder, interview, 25 March 2019.
41. David Dennis, interview, 23 April 2019.
42. Bob Crowder, 1964 Diary, 19 August 1964.
43. Ibid., 18 April, 31 May, 1 June, 25 July, 23 August 1964.
44. Bob Crowder, interview, 8 January 2019.
45. Madge Crowder to Bob Crowder, 25 April 1963; Madge Crowder to Bob Crowder, 20 March, 8 May 1963; Bill Crowder to Bob Crowder, 30 April 1963.
46. Bill Crowder to Bob Crowder, 30 April 1963.
47. www.audioculture.co.nz/people/mike-walker; Bob Crowder, 1964 Diary, 11 June 1964; 23 June 1964. Mike Walker came to dominate the jazz piano scene in New Zealand; the Mike Perjanik Band was the leading session group on the Auckland recording scene between 1964 and 1966: https://en.wikipedia.org/wiki/Mike_Perjanik
48. Ibid., 23 May, 27 June, 4 July 1964.
49. Ibid., 12 April, 13 June 1964.
50. Ibid., 10 May, 13 May, 7 June 1964.

51. http://adb.anu.edu.au/biography/mccarthy-john-keith-10910
52. Bob Crowder, 1964 Diary, 12 May 1964.
53. Ibid., 16, 25, 27, 28 May 1964.
54. Bob Crowder, interview, 8 January 2019.
55. Bob Crowder, pers. comm., 7 June 2023.
56. Ibid.
57. Ibid.
58. Bob Crowder, 1964 Diary, 20 June, 21 June 1964; Cash Account and Bill Book.
59. Interview, Bob Crowder, 8 January 2019.
60. Bob Crowder, 1964 Diary, 20 August 1964.
61. Bob Crowder to Madge and Bill Crowder, and Uncle Jack, 25 May 1966.
62. Bob Crowder, 1964 Diary, 24 July 1964; 30 July 1964.
63. Bob Crowder, interview, 8 January 2019.
64. Ibid.
65. Ibid.
66. Ibid.
67. Ibid.
68. R.A. Crowder, 'Vegetable with a difference: The witloof, widely grown in Europe, is being grown in Pukekohe', *New Zealand Journal of Agriculture*, vol. 111, no. 7, December 1965, p. 35.
69. Bob Crowder to Madge and Bill Crowder, and Uncle Jack, 14 February 1966.
70. Bob Crowder, interview, 25 March 2019.
71. Bob Crowder to Madge and Bill and Uncle Jack, 18 January 1966, 14 February 1966.
72. Bob Crowder, Diary, 8 April 1966.

3: LINCOLN AND THE HORTICULTURAL RESEARCH UNIT, 1966–76

1. Bob Crowder, Diary, 14 April 1966.
2. Bob Crowder to Madge & Bill Crowder, and Uncle Jack, 25 May 1966.
3. Bob Crowder, Diary, 22 April 1966; Bob Crowder to Madge & Bill Crowder, and Uncle Jack, 25 May 1966; Bob Crowder, Diary, 18, 25 June, 9, 16 July 1966, 6, 20, 21, 27, 28 August, 3, 4, 11 September, 8, 9, 15, 22 October 1966.
4. Interview, Bob Crowder, 21 May 2019.
5. https://livingheritage.lincoln.ac.nz/nodes/view/7635#idx47239; Bob Crowder, Diary, 24 December 1966.
6. It is possible he was directed to The Landing by his father in Devizes, as Fiona, the daughter of one of Bill's clients, had moved to Christchurch and was working there. According to Madge, Fiona's family 'are definitely well-to-do landowners, but like the majority of that type are definitely not "snobs"'. Madge Crowder to Bob Crowder, 4 June 1971. Bob Crowder, interview, 10 March 2020.
7. Bob Crowder, interview, 21 May 2019.
8. T.M. Morrison, 'University raises status of study in horticulture', *New Zealand Journal of Agriculture*, vol. 114, no. 1, January 1967, p. 58.

9. 'The new look in horticulture', *Lincoln College Magazine*, no. 94, 1969, p. 24: https://livingheritage.lincoln.ac.nz/nodes/view/1059#idx7394
10. Ibid.
11. Bob Crowder, Field Note Book, 1968. The fruit was grown by Graham Thiele; Bob Crowder, pers. comm., 28 July 2022; Bob Crowder, Field Note Book, 1968, 20 January 1969.
12. Bob Crowder, interview, 21 May 2019.
13. Bob Crowder, Weather Notes, 29 November 1968.
14. Bob Douglas, interview, 23 July 2019.
15. Bob Crowder to O.A. Lorenz (Dept. of Vegetable Crops, University of California, Davis), 10 October 1968; Bob Crowder to director, National Vegetable Research Station (Warwickshire, England), 30 April 1969.
16. C. Ward to Bob Crowder, 20 December 1968.
17. Bob Crowder to P. Perkinson, 10 October 1968.
18. R.C. Billman (Monsanto) to Bob Crowder, 18 April 1969.
19. Jack Garvin (commercial manager, Monsanto) to Bob Crowder, 16 June 1969.
20. Bob Crowder, interview, 21 May 2019.
21. Bob referred to the hailstorm in a letter to Stanhay Ltd's export sales manager: Bob Crowder to C. Ward, 14 March 1969.
22. Ibid.
23. R.A. Crowder, 'New techniques could boost vegetable growing', *New Zealand Journal of Agriculture*, vol. 118, no. 3, March 1969, p. 78.
24. Bob Crowder, interview, 21 May 2019.
25. US Tomato Statistics (USDA Economics, Statistics and Market Information System), 2010: https://usda.library.cornell.edu/concern/publications/br86b356q?locale=en
26. Ildi Carlisle-Cummins, 'From ketchup to California cuisine: How mechanical tomato harvesting prompted today's food movement' (2015): https://news.plantsciences.ucdavis.edu/2015/07/24/how-the-mechanical-tomato-harvester-prompted-the-food-movement/
27. Ibid.
28. Bob Crowder to Madge and Bill Crowder, 2 June 1969.
29. Bob Crowder, 'Report on short-term refresher leave taken from May–August 1969', Bob Crowder collection.
30. Bob Crowder to Madge and Bill Crowder, 2 June 1969.
31. Ibid.
32. Ian Thompson to Bob Crowder, 25 April 1974.
33. Bob Crowder, interview, 21 May 2019; Bob Douglas, interview, 23 July 2019.
34. R.C Billman (Monsanto) to Bob Crowder, 18 April 1969; Bob Crowder to Dr Hladik, 15 September 1969. Bob had met him at Lincoln; Bob Crowder, pers. comm., 28 July 2022; Bob Crowder to C.J. Ward, 15 September 1969; Bob Crowder to M. Shaul, 9 September 1969.
35. F. Lawson (South West Pacific marketing manager, Monsanto) to Bob Crowder, 2 May 1969; Bob Crowder to J.W. Garvin (Monsanto), 15 September 1969.
36. R.A. Crowder, *Development of Machine Methods for Extensive Crop Planting: Onions* (Lincoln College (University of Canterbury), 1968); R.A. Crowder, *Extensive Vegetable Production* (Lincoln College (University of Canterbury), 1968).

37. R.A. Crowder, 'Pelleted seed in onion production', *New Zealand Journal of Agriculture*, vol. 121, no. 5, November 1970, 85.
38. Ibid., 83. Note that mercurous chloride was never used in pelleting in Bob's trials.
39. Bob Crowder, interview, 21 May 2019.
40. For example, Bob Crowder, Field Note Book, 13 November 1968.
41. Bob Crowder, Field Notes, 23 August 1971, 11 October 1971.
42. Bob Crowder, Field Note Book, 15 October 1971.
43. Bob Crowder, interview, 21 May 2019.
44. Ibid.
45. Bob Crowder, Field Note Book, 3 and 30 December 1971.
46. Madge Crowder to Bob Crowder, 1 July 1971.
47. Ian Thompson to Bob Crowder, 21 July 1971.
48. Ian Thompson to Bob Crowder, 16 December 1971.
49. Ian Thompson to Bob Crowder, 11 May 1972.
50. Various letters and information sheets signed by Bob relate to this. See, for example, Bob Crowder, 'Christchurch Concerned Academics Committee' (undated but certainly June 1971), asking attendees of the initial meeting to contribute to the committee's expenses; W.E. Murphy (Victoria University of Wellington) to Bob Crowder, 16 July 1971 (accepting invitation to speak at a planned teach-in); David Thomson (Minister of Defence) to Bob Crowder, 12 July 1971; Hugh Templeton MP to Bob Crowder, 14 July 1971.
51. Minutes, Canterbury University Academic Staff Mobilisation, 30 July 1971, Bob Crowder collection. See also, for example, letter to colleagues on behalf of the Christchurch Area Concerned Academics Group about a planned 14 July mobilisation, 17 June 1972. Bob Crowder collection.
52. Bob Crowder to K. Buchanan, 9 July 1971. Buchanan was the founding professor of geography at Victoria University of Wellington, a position he held from 1953 to 1975.
53. Bob Crowder to Norman Kirk, 8 June 1972.
54. Madge Crowder to Bob Crowder, 16 August 1971.
55. Ibid.
56. Ibid.
57. Ibid.
58. Bill Crowder to Bob Crowder, 8 November 1971.
59. John Dickinson to Bob Crowder, 11 January 1972.
60. John Dickinson to Bob Crowder, 17 April 1972. This did in fact eventuate, and the Australian Labor Party won the federal elections for the first time in 23 years.
61. John Dickinson to Bob Crowder, 26 March 1972; Bob Crowder, pers. comm., 28 July 2022.
62. John Dickinson to Bob Crowder, 14 June 1973.
63. *Lincoln College Magazine*, 1972: https://livingheritage.lincoln.ac.nz/nodes/view/604#idx3642; https://livingheritage.lincoln.ac.nz/nodes/view/1850#idx13703. Bob Douglas confirmed that Gill left Lincoln around that time: Bob Douglas, interview, 23 July 2019; Bob Crowder, interview, 27 March 2023.
64. Bob Douglas, interview, 23 July 2019.
65. Bob Crowder, interview, 21 May 2019.
66. Bob Douglas, interview, 23 July 2019.

67. Bob Crowder, interview, 21 May 2019.
68. R.A. Crowder, 'Research into direct seeding of asparagus', *New Zealand Journal of Agriculture*, vol. 125, no. 1, July 1972: 45.
69. Bob Crowder, Field Note Book, undated but noted between entries in April and May 1972.
70. Bob Crowder, interview, 10 March 2020.
71. Bob Crowder, interview, 21 May 2019.
72. 'Drama', *Lincoln College Magazine*, 1972: https://livingheritage.lincoln.ac.nz/nodes/view/604#idx3629
73. Madge Crowder to Bob Crowder, 31 March 1972.
74. Madge Crowder to Bob Crowder, 4 October 1972.
75. Madge Crowder to Bob Crowder, 3 March 1973.
76. Bill Crowder to Bob Crowder, 20 February 1973.
77. Unknown, to Bob Crowder, 4 March 1970.
78. Ibid.
79. Bob Crowder, interview, 14 September 2021. In an earlier telling of this story, Bob said, 'The outstanding variety was given the name of "Crowder's Delight" by the garden party panel over the years of tastings.' Bob Crowder, 'Artichoke production, Lincoln College, 1969–1982' (unpublished, presumably 1982), 2.
80. Madge Crowder to Bob Crowder, 8 November 1977; interview, Bob Crowder, 14 September 2021.
81. Bob Crowder, interview, 14 September 2021.
82. Bob Crowder, in Scott McVarish, *The Greening of New Zealand: New Zealanders' visions of green alternatives* (Auckland: Random Century, 1992), p. 151.
83. Ian Thompson to Bob Crowder, November 1971.
84. Ian Thompson to Bob Crowder, 19 August 1971.
85. Ian Thompson to Bob Crowder, 11 May 1972.
86. Bob Crowder, interview, 10 March 2020; Bob Crowder, pers. comm., 28 July 2022; Bob Crowder, pers. comm., 7 June 2023.
87. Bob Crowder to Madge and Bill Crowder, 27 December 1972.
88. Bob Crowder to Madge and Bill Crowder, 25 January 1973.
89. Ibid.
90. Bob Crowder, pers. comm., 28 July 2022.
91. Bob Crowder to Madge and Bill Crowder, 25 January 1973; Peter Kopu to Bob Crowder, 11 March 1974.
92. Bob Crowder to Madge and Bill Crowder, 25 January 1973.
93. Peter Kopu to Bob Crowder, 12 November 1973; Peter Kopu to Bob Crowder, 6 December 1974; Peter Kopu to Bob Crowder, 24 February, 1975.
94. Bob Crowder, interview, 10 March 2020.
95. John Dickinson to Bob Crowder, 3 June 1974.
96. Ian Thompson to Bob Crowder, 21 September 1974.
97. Ibid.
98. Bob Crowder, Field Note Book, 26 October 1972; 8 November 1972.
99. R.A. Crowder, 'Onion production: A comparison between yield and grading', *New Zealand Journal of Agriculture*, vol. 127, no. 5, November 1973, pp. 26–27.

100. Ian Thompson to Bob Crowder, 19 July 1973.
101. Madge Crowder to Bob Crowder, 8 April 1973.
102. Madge Crowder to Bob Crowder, 28 June 1973.
103. Ian Thompson to Bob Crowder, 19 July 1973.
104. https://livingheritage.lincoln.ac.nz/nodes/view/1122#idx8066
105. Untitled report on 1973 California trip, 17 October 1973.
106. Bob Douglas, interview, 23 July 2019.
107. Ibid.
108. Ibid.
109. Madge Crowder to Bob Crowder, 5 February 1974.
110. Bob Crowder to Madge and Bill Crowder, 20 October 1974.
111. Madge Crowder to Bob Crowder, 27 March 1974; https://livingheritage.lincoln.ac.nz/nodes/view/1850#idx13703
112. R.A. Crowder, 'Mechanical harvesting of tomatoes', *New Zealand Journal of Agriculture*, vol. 129, no. 5, November 1974, p. 46.
113. Bill Crowder to Bob Crowder, 3 March 1975; 12 March 1975.
114. Madge Crowder to Bob Crowder, 26 March 1975; 23 April 1975.
115. H.G. Hunt to Bob Crowder, 29 May 1975; Personnel officer to Bob Crowder, 1 July 1975.
116. Madge Crowder to Bob Crowder, 11 June 1975.
117. The reference is to Anthony Neil Wedgwood Benn, British politician and member of Cabinet in the Labour governments of the 1960s and 70s; Madge Crowder to Bob Crowder, 7 May 1975.

4: ORGANIC CONVERSION, 1975–83

1. Bob Crowder, 'One year away from Lincoln: A general report to council on how an exchange year was spent at Bath University in Britain', 21 September 1976, Bob Crowder collection.
2. www.walcotstreet.com/archive
3. Bob Crowder Notebook, undated (but almost certainly 1975. Entries later in the notebook, however, are from 1977–79).
4. Bob Crowder, interview, 10 March 2020.
5. Bob Crowder, interview, 10 March 2020; Bill Crowder to Bob Crowder, 17 September 1976.
6. Bill Crowder to Bob Crowder, 18 August 1976.
7. Bob Crowder to Madge and Bill Crowder, 10 August 1976.
8. Ibid.
9. Bob Crowder, Notebook, November 1976 and 29 March 1977.
10. Press cutting, untitled, undated (but a note indicates October 1930), Bob Crowder Collection; Erewhon Morris Dancers, 'Statement of policy regarding the Australian Morris Ring,' undated (but presumed to be 1978), Bob Crowder Collection.
11. Bob Crowder, 'Erewhon Morris Report April 1978', Bob Crowder Collection.
12. When he felt Erewhon was strong enough, he argued firmly against a rule emerging in Australasia that Morris should be for men only. 'We follow the belief that "Folk" is a living

tradition for the people of the times. In Christchurch there are not enough men interested in traditional male Morris dancing to maintain a side. There are sufficient persons in Christchurch very interested in joining a living tradition of Morris dancing that involves mixed teams. The Erewhon Morris is a lively and very social side that would fall to pieces if any attempt at segregation was made.' Erewhon Morris Dancers, 'Statement of policy regarding the Australian Morris Ring,' undated (but presumed to be 1978), Bob Crowder Collection.
13. Press cutting, untitled, undated, Bob Crowder Collection.
14. Judith McArthur, 'Have bells … will travel,' *Straight Furrow*, 6 July 1978, p. 18.
15. Erewhon Morris, 'Morris dancing tradition comes to Christchurch', undated. Bob Crowder Collection.
16. Bob Crowder, 'Erewhon Morris report April 1978', Bob Crowder Collection.
17. *Central Otago News*, 5 January 1978. Bob recalled they were usually hosted by well-known restaurateur Fleur Sullivan, then living at Clyde. The group camped in a cherry orchard and participated in, if not started, the ritual of the cherry stone spitting competition. Bob Crowder, pers. comm., 28 July 2022; *The Dominion*, 3 June 1978; '50th anniversary pageant big draw', *The Press*, 17 November 1979; *Otago Daily Times*, 31 December 1979.
18. 'Public appearances of the Erewhon Morris Dancers, April 1979–1980', appended to the 'Erewhon Morris report, 15 April 1980', Bob Crowder Collection.
19. Madge Crowder to Bob Crowder, 6 December 1976.
20. Madge Crowder to Bob Crowder, 2 August 1977.
21. Bob Crowder to Leonard Broadbent, 1 October 1976.
22. Bob Crowder, interview, 7 February 2022.
23. Bob Crowder, interview, 14 January 2020.
24. Haikai Tane, email to author, 4 May 2020.
25. Tim Porteous, interview, 21 September 2022.
26. Haikai Tane, email to author, 4 May 2020.
27. Haikai Tane, 'Ecogenesis of the BHU', undated article manuscript, based on notes for a lecture given between 2000 and 2002. Courtesy Haikai Tane.
28. Tim Maples, interview, 3 March 2020.
29. Tim Maples, interview, 3 March 2020. Albert Howard's book is an undisputed classic of organic farming literature. Eve Balfour's book, published just three years later, is of equal standing. Balfour later founded the British Soil Association (in 1946). See Albert Howard, *An Agricultural Testament* (London: Oxford University Press, 1940); Eve Balfour, *The Living Soil and the Haughley Experiment* (London: Faber & Faber, 1943).
30. Haikai Tane, email to author, 4 May 2020.
31. Bob Crowder, interview, 14 January 2020; Tim Maples, interview, 3 March 2020.
32. Bob Crowder, interview, 10 March 2020.
33. Bob Crowder, 'Education in organics: Practising that which is preached', (undated but likely 1996. Probably presented at IFOAM), p. 3.
34. Bob Crowder, undated, untitled reference for Tim Maples. This letter was found in a folder of papers from the period 1980–82: Small Farmers & Organics Folder, Bob Crowder Collection.
35. Bob Crowder to Tony West, 15 September 1980. Note that Tim Maples was never officially on the teaching staff at Lincoln.

36. Mollie Chalklen was assistant editor from 1979, so Haikai's time can only have been two years at most.
37. Haikai Tane, 'The Soil Association of New Zealand: Conference 1977', in *Soil & Health*, August/September 1977, p. 13.
38. Bob Crowder, interview, 14 January 2020.
39. A story emerged in later years that the organic unit was birthed in 1976, but this cannot be correct as Bob later often referred to the impetus of Tim Maples in getting the organic work underway, and Tim was still in India in 1976. See, for example, Bob Crowder, 'Biological Husbandry Unit – Lincoln University', (c.1993), article manuscript. Bob Crowder Collection; 'BHU Organic Education and Training – Organic Training College of Aotearoa New Zealand Business Plan', undated (but certainly 2005. The business plan in question was discussed at a meeting of the trust board and outlined in the 'Manager's report to the BHU Board of Trustees', 7 November 2005); BHU Organics Trust Board AGM Minutes, 20 November 2012; 'BHU Review and Strategy Update', October 2021, BHU Organics Trust Archive; Tim Maples, interview, 3 March 2020.
40. 'Horticulture Club', *Lincoln College Magazine*, 1977, p. 25: https://livingheritage.lincoln.ac.nz/nodes/view/1124#idx8307
41. Tim Maples, interview, 3 March 2020.
42. Jo Blakely, interview, 18 October 2021.
43. Ibid.
44. Tim Maples, interview, 3 March 2020.
45. Bob Crowder, interview, 10 March 2020.
46. Gregory Barton, *The Global History of Organic Farming* (Oxford: Oxford University Press, 2018), pp. 193–94.
47. Roland Chevriot, quoted in Barton, *The Global History of Organic Farming*, pp. 192.
48. Barton, *The Global History of Organic Farming*, p. 192.
49. S. Padel and N. Lampkin, 'The development of governmental support for organic farming in Europe', in W. Lockeretz (ed.), *Organic Farming: An international history* (Oxfordshire: CAB International, 2007), p. 96.
50. W. Lockeretz, 'What explains the rise of organic farming?', in Lockeretz (ed.), *Organic Farming*, p. 2.
51. Bergland report, quoted in Lockeretz, 'What explains the rise of organic farming?', p. 2.
52. Jill died in the 1980s and was buried under an oak tree at Lincoln, marked with a stone that was later relocated to Bob's garden at Ashgrove Terrace. 'Poor Jill … She was still very bright and alert when she died … That was more an emotional upset than my mother dying.' Bob Crowder, interview, 18 October 2021; Bob Crowder, pers. comm., 28 July 2022.
53. Bob Crowder, interview, 14 September 2021.
54. Ibid.
55. Bill Crowder to Bob Crowder, 21 September 1980.
56. Madge commented on this renewed happiness in a letter: Madge Crowder to Bob Crowder, 13 October 1980.
57. Bob Crowder, quoted in Sandra Stewart, 'Lincoln College students are taking the mystery out of "muck"', *The Star*, 20 March 1982, p. 5.
58. Bob Crowder, interview, 7 February 2022.
59. Ibid.

60. Bob Crowder to Tony West, 23 October 1980.
61. Bob Crowder to IFOAM, 'Nature et Progrès', 9 September 1981; Bob Crowder to Ken Ettlinger (Peconic Seed and Plant Co-operative, Manorville), 6 January 1982.
62. Bob Crowder to Mr and Mrs D.R. Blair, 16 September 1981.
63. Peter Waugh (Federated Farmers) to Bob Crowder, 5 February 1981.
64. Bob Crowder to Peter Waugh, 23 February 1981.
65. Ibid.
66. Ibid.
67. R.D. McLagan to Bob Crowder, 23 April 1981.
68. Bob Crowder, interview, 7 February 2022; See Bob Crowder to Dr R. Parnes (Woods End Agricultural Institute, Maine), 16 September 1981.
69. The list of places Bob visited in 1981 is compiled from multiple sources, including his request for leave in 1984, which listed sites visited in 1981: 'Application for Sabbatical Leave 1984', Bob Crowder collection.
70. Bob Crowder, interview, 7 February 2022; Bob Crowder to Dr R. Parnes (Woods End Agricultural Institute, Maine), 16 September 1981.
71. Bob Crowder to IFOAM, 'Nature et Progrès, 9 September 1981.
72. Tim Jenkins to Matt Morris, email, 14 November 2021.
73. 'Farming observations: Biological farming at Haughley Research farms', *Soil & Health*, August/September 1979, p. 7; 'Biological farming', *Soil & Health*, October/November 1980, pp. 17–22; Jack Meechin, 'How organic or biological farming compares with chemical farming', *Soil & Health*, June/July 1981, p. 25.
74. 'The Tauranga Community College course in biological husbandry, lecture no. 7. "Climate factors"', *Soil & Health*, Summer 1982, pp. 14–16; 'What's in a name?', *Soil & Health*, February/March 1982, p. 4.
75. 'The 39th annual meeting of the Soil Association', *Soil & Health*, June/July 1981, p. 37.
76. Exactly when Mollie Chalklen became involved with the association is not known. She is not listed among national council members in Jack Whitelaw's 1976–77 Notebook. She is, however, referred to in a Canterbury branch report from 1979 in which she had raised $200 through a plant stall in Cathedral Square. 'Canterbury', *Soil & Health*, May/June 1979, p. 36.
77. Mark Vette to Bob Crowder, 18 May 1981.
78. Chris May, interview, 19 May 2020.
79. Ibid.
80. Chis May to author, 26 August 2022.
81. Federated Farmers of New Zealand Inc, 'Seminar on biological farm management techniques' (seminar programme), Small Farmers & Organics Folder, Bob Crowder Collection; 'Never mind the philosophy: Feel the wallet', *Straight Furrow*, 'Organic Farming Supplement', 11 June 1982, p. 2.
82. Federated Farmers of New Zealand Inc, 'Seminar on biological farm management techniques'.
83. Mollie Chalklen, 'National president's report to the annual conference 1983', Organic Growing, Winter 1983, p. 3; Federated Farmers of New Zealand Inc, 'Seminar on biological farm management techniques' (seminar programme), Small Farmers & Organics Folder, Bob Crowder Collection.

84. Federated Farmers of New Zealand Inc, 'Seminar on biological farm management techniques'; Bob Crowder, 'Can we think big – naturally?', Straight Furrow, 'Organic Farming Supplement', 11 June 1982, pp. 8–9.
85. Ruth Richardson, as the legal advisor to Federated Farmers, was instrumental in this development.
86. Peter Waugh, interview, 22 June 2020.
87. Lockeretz, 'What explains the rise of organic farming?', p. 1.
88. Peter Waugh, interview, 22 June 2020; Crowder, 'Can we think big naturally?', pp. 8–9. Bob recalled that the supplement included an article on Maxicrop, a seaweed-based fertiliser that was considered controversial at the time. Bob Crowder, pers. comm., 28 July 2022.
89. Crowder, 'Can we think big naturally?', p. 9.
90. Chris May, interview, 19 May 2020.
91. Dave Woods (Doubleday Research Association of New Zealand), to various, 28 July 1982; James Piper to Dave Woods, 27 July 1982. Bob does recall being there, although this is not explicit in the record: Bob Crowder, pers. comm., 28 July 2022. Chris May also noted that they used some of their $500 grant to get Bob to a meeting in Auckland, although this may have been the subsequent Pitt Street meeting. Chris May to author, 26 August 2022.
92. Chris May, interview, 19 May 2020.
93. Jenny May, interview, 19 May 2020.
94. Ibid.
95. Interview, Bob Crowder, 18 October 2021.
96. Peter Waugh to Bob Crowder, 2 August 1982.
97. For example, see Georgia Shearer, Daniel H. Kohl, Diane Wanner, George Kuepper, Susan Sweeney and William Lockeretz, 'Crop production costs and returns on Midwestern organic farms: 1977 and 1978', *American Journal of Agricultural Economics*, vol. 63, no. 2, 1981, pp. 264–96. Bob cited Lockeretz in particular, as he had seen him speak on this topic.
98. Bob Crowder to Peter Waugh, 12 August 1982.
99. Ibid.; Bob Crowder, pers. comm., 28 July 2022.
100. Mollie Chalklen, 'From the president', *Organic Growing*, Autumn 1983, p. 2.
101. Richard Hudson and John Calvert (eds), *Ecological Agriculture: Review and annotated bibliography* (Wellington: Federated Farmers of New Zealand (Inc), 1983), p. 1.
102. Chalklen, 'From the president'.
103. 'Annual conference 1983', *Organic Growing*, Autumn 1983, p. 6.
104. Chris May, 'From the president', *Organic Growing*, Winter 1983, p. 2.
105. Mollie Chalklen, 'National president's report to the annual conference 1983', *Organic Growing*, Winter 1983, p. 3.
106. 'Conference speakers', *Soil & Health*, Winter 1983, p. 6.
107. R.A. Crowder, 'A guide to organic growing profitability', *Soil & Health*, Spring 1983, pp. 12–14.
108. *Soil & Health*, Autumn 1983, pp. 5, 6.
109. 'Lincoln College man's view', *Soil & Health*, Winter 1983, p. 23.
110. Interview, Bob Crowder, 14 January 2020.
111. Bob Crowder, 'Field Notes', undated, but other references in the notebook show that this visit to Scott's farm was no earlier than 1981 and no later than August 1983.
112. 'Lincoln College man's view', *Soil & Health*, Winter 1983, p. 23.

113. Advertisement, *Soil & Health*, August/September 1979, p. 26; Peter Waugh, interview, 22 June 2020.
114. 'Lincoln College man's view', *Soil & Health*, Winter 1983, p. 23.
115. 'Our branches report', *Soil & Health*, Summer 1983, p. 44.
116. Bob Crowder, 'Think big – naturally', *Soil & Health*, Autumn 1984, p. 18.
117. *Soil & Health*, Autumn 1984.

5: DEVELOPING ORGANICS, 1984–87

1. https://teara.govt.nz/en/photograph/40380/attic-coffee-house-christchurch-around-1957
2. Bob Crowder, interview, 27 August 2018.
3. Madge Crowder to Bob Crowder, 19 June 1978; Madge Crowder to Bob Crowder, 17 September 1979; Madge Crowder to Bob Crowder, 29 June 1980.
4. Madge Crowder to Bob Crowder, 2 May 1982; Bill Crowder to Bob Crowder, 11 March 1983; Bob Crowder, interview, 27 August 2018.
5. Valerie Thompson, 'Report on biological farming', *Soil & Health*, Autumn 1984, p. 32.
6. Perry Spiller, 'Organics in New Zealand over the last decade', manuscript c. 1992, Bob Crowder Collection.
7. Bob Crowder, 'Biological husbandry: Which way are we steering?', *Soil & Health*, Autumn 1984, p. 10.
8. Crowder, 'Biological Husbandry, p. 10.
9. Ibid.
10. Chris May, 'From the president', *Soil & Health*, Autumn 1984, p. 2.
11. 'Organic market gardening examined', *Soil & Health*, Winter 1984, p. 17.
12. Chris May, '"You can change the world": Project GRO', *Soil & Health*, Winter 1984, p. 2.
13. Chris May, 'President's report 1984', *Soil & Health*, Winter 1984, p. 4.
14. Bob Crowder, 'Establishment of Biological Husbandry Demonstration Unit', *Soil & Health*, Spring 1984, p. 2.
15. Crowder, 'Establishment of Biological Husbandry Demonstration Unit', p. 3.
16. Bob Crowder, interview, 2 June 2020; Perry Spiller, 'Perry's Page', *Soil & Health*, Summer 1986, p. 46.
17. Bob Crowder, interview, 2 June 2020.
18. Bob Douglas, interview, 23 July 2019.
19. Ibid.
20. Brendan Hoare, interview, 4 June 2020.
21. Ibid.
22. *Canterbury Organic Producers Newsletter*, October 1986; *Canterbury Organic Producers Newsletter*, October 1987; *Canterbury Organic Producers Newsletter*, June 1990.
23. Joanne Blakely, 'Personal report', *Soil & Health*, Autumn 1987, p. 31.
24. Jo Blakely, interview, 18 October 2021.
25. Jon Manhire, interview, 21 October 2021.
26. Jon Manhire to Matt Morris, 17 February 2021.
27. Jon Manhire, interview, 21 October 2021.
28. Bob Crowder, 'Biological Husbandries', *Soil & Health*, Spring 1984, p. 36.

29. Crowder, 'Biological Husbandries', p. 37.
30. Chris May, 'The president's piece', *Soil & Health*, Autumn 1985, p. 2.
31. Geoff Barnett, interview, 14 September 2021.
32. Bob Crowder, interview, 10 March 2020; Bob Crowder, interview, 7 February 2022.
33. Bob Crowder, 'The "organic" movement is becoming respectable', *Soil & Health*, Winter 1985, p. 11.
34. Crowder, 'The "organic" movement is becoming respectable', p. 11.
35. Bob Crowder, interview, 2 June 2020.
36. Crowder, 'The "organic" movement is becoming respectable', p. 11.
37. Ibid.
38. Bob had heard Hardy Vogtmann speak at the XXth International Horticultural Congress held in Sydney, 1978. See W.J. Greenhalgh, 'Press statement: Congress to discuss vegetable growing … from the equator to the poles', 20 December 1977. On Youngberg, see Gregory Barton, *The Global History of Organic Farming* (Oxford: Oxford University Press, 2018), pp. 188–90.
39. Bob Crowder, 'IFOAM 1984, 1986', Diary, undated diary entry (August 1984).
40. Bob Crowder, 'IFOAM 1984, 1986', Diary, 15 August 1984.
41. Crowder, 'The "organic" movement is becoming respectable', p. 13.
42. Holger Kahl, diary entry, August 1984. Reprinted in 'Snippets from a Diary' (presumably in 1999).
43. Hardy Vogtmann, 'Sustainable agriculture: Not just desirable, vital', *Soil & Health*, Autumn 1985, p. 20.
44. Bernward Geier, 'University research in organics growing', *Soil & Health*, Autumn 1986, pp. 17–19.
45. Barton, *The Global History of Organic Farming*, p. 192; David Sayers, 'Matahui – an organic kiwifruit orchard', *Soil & Health*, Summer 1984–85, p. 9.
46. *Soil & Health*, Winter 1985, back cover.
47. Jonathon Toye, 'It's not easy for produce to earn a Bio Gro label', *Soil & Health*, Summer 1985–86, p. 28.
48. 'Major step in furthering aims', *Soil & Health*, Winter 1985, p. 2.
49. Ibid.
50. Chris May, interview, 19 May 2020.
51. Plant variety rights morphed into a discussion about genetic engineering by mid-1986. See Perry Spiller, 'Perry's Page', *Soil & Health*, Winter 1986, p. 18; Perry Spiller, 'From the education officer's notebook', *Soil & Health*, Summer 1985–86, p. 21.
52. New Zealand Biological Producers Council, Inc., Minutes of 3rd Annual General Meeting to the Members of the Council, 15 March 1986; Minutes of the Annual General Meeting of the New Zealand Biological Producers Council, 27 March 1988, Bob Crowder Collection.
53. Bob Crowder to Colin Moyle, 12 February 1985.
54. Colin Moyle to Bob Crowder, 27 February 1985.
55. Bob Crowder to Colin Moyle, 21 June 1985.
56. Colin Moyle to Bob Crowder, 12 July 1985.
57. Bob Crowder to Colin Moyle, 9 August 1985.
58. Colin Moyle to Bob Crowder, 5 September 1985.

59. Bob Crowder to Mike Moore, 21 June 1985; Bob Crowder to Simon Upton, 21 June 1985; Simon Upton to Bob Crowder, 31 July 1985.
60. Perry Spiller, 'From the education officer's notebook', *Soil & Health*, Summer 1985–86, p. 21.
61. Ken Chappell to chair of Whangarei Branch Soil Association, 25 March 1986. Bob Crowder Collection.
62. New Zealand Biological Producers Council Inc, Minutes of Executive Committee Meeting, 12 April 1986. Bob Crowder Collection.
63. New Zealand Biological Producers Council Inc., Minutes of Executive Committee Meeting, 14 May 1986. Bob Crowder collection.
64. Ibid.
65. *Canterbury Organic Producers Newsletter*, October 1986.
66. New Zealand Biological Producers Council Inc., Minutes of Executive Committee Meeting, 6 December 1986. Bob Crowder collection.
67. David Sayers to Chris May, 21 May 1986. Bob Crowder collection.
68. New Zealand Biological Producers Council Inc., Minutes of Executive Committee Meeting, 27 June 1986. Bob Crowder collection.
69. Bruce Collins to Grant Mitchell, 2 August 1986. Bob Crowder collection.
70. New Zealand Biological Producers Council Inc., Minutes of Executive Committee Meeting, 30 August 1986. Bob Crowder collection.
71. Bob Crowder, interview, 2 June 2020.
72. Perry Spiller to author, 24 August 2022; Chris May, interview, 19 May 2020; Bob Crowder, interview, 2 June 2020.
73. Robin Scott to the secretary, NZBPC, 4 July 1986, appended to New Zealand Biological Producers Council Inc, Minutes of Executive Committee Meeting, 27 June 1986. Bob Crowder collection.
74. Marinus La Rooj, quoted by Perry Spiller, 'Meeting with the Ministry of Agriculture and Fisheries, Wellington – 15 August 1986', in New Zealand Biological Producers Council Inc, Minutes of Executive Committee Meeting, 30 August 1986. Bob Crowder collection.
75. Ibid.
76. Bob Crowder, paraphrased in New Zealand Biological Producers Council Inc, Minutes of Executive Committee Meeting, 6 December 1986. Bob Crowder collection.
77. *Canterbury Organic Producers Newsletter*, August 1988.
78. David Musgrave, interview, 3 January 2022; *Canterbury Organic Producers Newsletter*, June 1990.
79. Bob Crowder, 'IFOAM 1984, 1986', Diary, 12 August 1986.
80. New Zealand Biological Producers Council Inc, Minutes of General Executive Committee Meeting, 27 June 1986. Bob Crowder collection.
81. Bob Crowder, interview, 10 March 2020.
82. Perry Spiller, 'Perry's Page', *Soil & Health*, Spring 1986, p. 37.
83. Bob Crowder, 'Santa Cruz conference', *Soil & Health*, Autumn 1987, p. 30.
84. Ibid.
85. Ibid.
86. Brain Chamberlain, 'Opening address', *Soil & Health*, Winter 1987, p. 12.
87. Robin Scott, 'Policy and prospects', *Soil & Health*, Winter 1987, p. 14.

88. Hardy Vogtmann, quoted in Bob Crowder, 'European organics', *Soil & Health*, Winter 1987, p. 26.
89. Robin Scott, 'Policy and prospects', *Soil & Health*, Winter 1987, p. 14.
90. Ibid.
91. Ibid., p. 15.
92. Anthony Haystead, 'Organic farming survey', *Soil & Health*, Spring 1987, p. 8.
93. Perry Spiller, interview, 9 June 2020.
94. Haystead, 'Organic Farming Survey', p. 9.
95. Ibid., p. 12.
96. 'MAF – BPC Standards Discussion', appended to New Zealand Biological Producers Council Inc, Minutes of the Executive Committee meeting, 7 November 1987. Bob Crowder Collection.
97. Bob Crowder, pers. comm., 13 June 2020.
98. 'MAF – BPC Standards Discussion', appended to New Zealand Biological Producers Council Inc Minutes of the Executive Committee meeting, 7 November 1987. Bob Crowder Collection.
99. 'Health News: On the conference', *Soil & Health*, Spring, 1986, p. 4. Even in 1987 it was still called the Biological Husbandry Demonstration Unit, as in 'Flame weeding for Lincoln?', *ProGro News*, June 1987, p. 3.
100. 'Organics – Key to Health: 44th Annual Conference, Soil Association of NZ, Inc', *Soil & Health*, Autumn 1986, p. 24.
101. Bob Crowder, pers. comm., 28 July 2022.
102. Bob Crowder, 'Biological Husbandry Unit – Lincoln University' (undated but likely 1993 or 1994).
103. Bob Crowder, 'Education in organics: Practising that which is preached' (undated, but likely 1996. Probably presented at IFOAM), p. 6.
104. Ibid.
105. Bob Crowder, 'Biological Husbandry Unit – Lincoln University' (undated but likely 1993 or 1994), Bob Crowder collection.
106. Bob Crowder, 'Report on the Progress of the Biological Husbandry Demonstration Unit 1986–87 Season', Bob Crowder collection.
107. Bob Crowder, 'Biological Husbandry Unit – Lincoln University' (undated but likely 1993 or 1994).
108. Perry Spiller, 'Biological producers' council news', *Soil & Health*, Summer 1987, p. 40.
109. Brendan Hoare, interview, 4 June 2020.

6: NEW ZEALAND ORGANICS ON THE WORLD STAGE, 1988–92

1. New Zealand Biological Producers Council Inc, Minutes of the Annual General Meeting, 27 March 1988, Bob Crowder collection; New Zealand Biological Producers Council Inc, Minutes of the executive committee meeting, 30 July 1988, Bob Crowder collection.
2. Inspection fees rose from $2118.35 in 1985 to $3510.70 in 1986, and to $9202.73 in 1987. New Zealand Biological Producers Council Inc, 'Income and expenditure account

for nine months ended 31 December 1985'; 'Income and expenditure account for the year ended 31 December 1986'; 'Income and expenditure account for the year ended 31 December 1987', Bob Crowder collection; New Zealand Biological Producers Council Inc, Minutes of the executive committee meeting, 7 November 1987, Bob Crowder collection.

3. Bob Crowder to Bruce and Denize McGill, 5 September 1989.
4. Bob Crowder, interview, 2 June 2020; Bob Crowder, interview, 10 March 2020.
5. Sandy Fritz to Bob Crowder, 24 June 1987.
6. Both Perry Spiller and Bob Crowder commented that they often joked together about the European pronunciation of IFOAM as EFOAM – and that the 'E' really stood for European rather than 'I' for International.
7. New Zealand Market Board Development Act (1986 no. 1): www.nzlii.org/nz/legis/hist_act/nzmdba19861986n1360/
8. New Zealand Biological Producers Council Inc, Minutes of the executive committee meeting, 26 March 1988, Bob Crowder collection.
9. Bob Crowder to Sandy Fritz, 17 February 1988.
10. Sandy Fritz to Bob Crowder, 25 February 1988.
11. Bob Crowder, 'New Zealand – Pure and Natural – Myth or Reality', article manuscript, undated (but certainly 1988 as it references the January 1989 IFOAM World Board meeting as upcoming). Bob Crowder collection.
12. Perry Spiller, email to author, 5 September 2022.
13. Bob Crowder to Sandy Fritz, 13 April 1988.
14. Sandy Fritz to Perry Spiller, 29 June 1988.
15. Perry Spiller to Sandy Fritz, 8 June 1988; Sandy Fritz to Perry Spiller, 24 January 1989; Minutes of the Biological Producers Council, 10 February 1990; Minutes of the executive committee of the Biological Producers Council, 9 and 10 February 1991.
16. Perry Spiller to author, 24 August 2022.
17. Minutes of the Biological Producers Council, 27 May 1989.
18. Bob Crowder, 'The New Zealand Organic Standard', article manuscript, February 1989. Bob did not attend this conference, but was proposed for the board by Nic Lampkin.
19. New Zealand Biological Producers Council Inc, President's report, Minutes of the Annual General Meeting, 12 March 1989, Bob Crowder collection.
20. Judith Cormick to Naturkost, 28 July 1989.
21. Bernward Geier to Bob Crowder, 22 August 1989.
22. Bob Crowder to Bernward Geier, 7 September 1989.
23. Jan von Ledebur to Perry Spiller, 26 July 1989.
24. Ibid.
25. Bob Crowder to Jan von Ledebur, 6 September 1989; New Zealand Biological Producers Council Inc, Minutes, 10 February 1990, Bob Crowder collection.
26. New Zealand Biological Producers Council Inc, President's report, Minutes of the Annual General Meeting, 31 March 1990.
27. O. Schmid, 'Development of standards for organic farming', in W. Lockeretz (ed.), *Organic Farming: An international history* (Oxfordshire: CAB International, 2007), p. 170.
28. E. Wynen and Sandy Fritz, 'NASAA and organic agriculture in Australia', in Lockeretz, *Organic Farming*, p. 232. Note their claim that this occurred in 1994, which does not concur with Schmid's chronology.

29. Bob Crowder, interview, 2 June 2020.
30. Bob Crowder, Confidential report to NZBPC (and Bernward Geier), Executive Meeting of NASAA, 23–24 September 1989.
31. Bob Crowder, Confidential Report to NZBPC (and Bernward Geier), Executive Meeting of NASAA, 23–24 September 1989.
32. Perry Spiller, 'Symposium snapshots', in *Soil & Health*, Winter, 1988, p. 18; Biological Husbandry Unit, Publicity Income Report: 17 August 1995 (a report on the 1994 IFOAM conference which also references the 1988 conference), Bob Crowder collection; New Zealand Biological Producers Council Inc, Minutes of executive committee meeting, 11 June 1988, Bob Crowder collection.
33. Bob Crowder, 'Project GRO achieves its AIM', article manuscript, August 1988, Bob Crowder collection.
34. Gillian Woods, 'Symposium snapshots', *Soil & Health*, Winter, 1988, p. 18.
35. Ibid.
36. Bernward Geier, quoted in Gillian Woods, 'Symposium snapshots', *Soil & Health*, Winter, 1988, p. 22.
37. Ibid.
38. Gillian Woods, 'Symposium snapshots', *Soil & Health*, Winter, 1988, p. 22.
39. Brendan Hoare, interview, 4 June 2020.
40. Ibid.
41. New Zealand Biological Producers Council Inc, Minutes of the executive committee meeting, 11 February 1989, Bob Crowder collection.
42. Kim Stevenson to Bernward Gaier, fax transmission date: 16 June 1989.
43. Ibid.
44. Kim Stevenson to Bernward Geier, 26 July 1989.
45. Ibid.
46. Bernward Geier to Bob Crowder, 26 July 1989.
47. Bob Crowder to Bernward Geier, 17 August 1989.
48. Bob Crowder to Bernward Geier, 7 September 1989.
49. Bob Crowder to Ian Cornforth, 11 August 1989.
50. Ibid.
51. Bob Crowder, Confidential Report to NZBPC (and Bernward Geier), Executive Meeting of NASAA, 23–24 September 1989.
52. New Zealand Biological Producers Council Inc, Minutes of the Annual General Meeting, 31 March 1990, Bob Crowder collection.
53. Ibid.
54. Minutes of the New Zealand Biological Producers Council Inc, 16 May 1990.
55. Minutes of the New Zealand Biological Producers Council Inc, 29 and 30 September 1990.
56. Minutes of the New Zealand Biological Producers Council Inc, 24 and 25 November 1990.
57. Minutes of the New Zealand Biological Producers Council Inc, 9 and 10 February 1991.
58. Perry Spiller, 'Organics in New Zealand over the last decade', article manuscript c. 1992, Bob Crowder collection.
59. Minutes of the New Zealand Biological Producers Council Inc, 14 May 1991, Bob Crowder collection.

60. Bob Crowder to Vicki Buck, 11 January 1990.
61. Bob Crowder to Diana Shand, 9 April 1990; Bob Crowder to Diana Shand, 19 July 1990. The politics of this taxed Bob for several months, as seen in letters between Bob and Gerry Butler, president of the Soil Association of South Australia. In them, Butler outlined his plans to put in a bid to host IFOAM, and Bob's frustration with this is evident. Not only could this endanger his bid, but perceived fragmentation of the Australasian organic movement could be potentially damaging for the whole discussion around regional autonomy within IFOAM. See, for example, Gerry Butler to Bob Crowder, 17 January 1990; Bob Crowder to Gerry Butler, 28 February 1990.
62. Bob Crowder to Diana Shand, 9 April 1990; John Longworth (DSIR director) to Bob Crowder, 27 June 1990.
63. Bob Crowder to IFOAM World Board members, 19 September 1990.
64. Minutes of the Executive Committee of the Biological Producers Council In, 14 May 1991; Minutes of the Executive Committee of the Biological Producers Council, 7 February 1992, Bob Crowder collection.
65. Bob Crowder, Biological Producers Council President's Report, 15 May 1991, Bob Crowder collection.
66. Bob Crowder to Bernward Geier, 19, 21 September 1992.
67. A travel fund was proposed. Minutes of the Executive Committee of the Biological Producers Council, 8 February 1992, Bob Crowder collection.
68. Bob Crowder, Biological Producers Council President's Report, 15 May 1991, Bob Crowder collection.
69. Bob Crowder to Richard Cox, 3 April 1991.
70. Bob Crowder to Dieter Proebst, 29 April 1992.
71. Bob Crowder, pers. comm., 28 July 2022.
72. Bob Crowder, Biological Producers Council President's Report, 15 May 1991, Bob Crowder collection.
73. Bob Crowder, 'Report on Biological Husbandry Unit, Lincoln University 1989–90 Season' (1990, apparently for Project GRO), p. 3.
74. Ibid., p. 2.
75. See, for example, Bob Crowder, 'Biological Husbandry Unit – Lincoln University', (c. 1993) article manuscript, p. 6, Bob Crowder collection.
76. Bob Crowder, quoted in Scott McVarish, *The Greening of New Zealand: New Zealanders' Visions of Green Alternatives* (Auckland: Random Century, 1992), p. 164.
77. Sol Morgan, interview, 10 March 2020.
78. Ibid.
79. See, for example, Bob Crowder to R.A. Martin (NZ Kiwifruit Marketing Board), 1 July 1991.
80. See folder 'I.f.o.a.m. Corporate Membership/Conference 1994', Letters and Responses, Bob Crowder collection.
81. Professor Richard Rowe to Michael Gaffany (NZ Fruitgrowers Federation), 15 August 1991.
82. Bob Crowder to Max Lilley, 10 July 1992.
83. Max Lilley to Bob Crowder, 22 July 1992.
84. Bob Crowder to Max Lilley, 28 July 1992.

85. Max Lilley, 'From the President', *Commercial Grower*, May/June 1992, p. 3.
86. Bob Crowder to Max Lilley, 10 July 1992; Max Lilley to Bob Crowder, 22 July 1992; Bob Crowder to Max Lilley, 28 July 1992.
87. Bob Crowder to Max Lilley, 10 July 1992.
88. Peter Silcock to Bob Crowder, 27 August 1992.
89. Certificate of BioGro Licence to Bob Crowder of the Biological Husbandry Unit, #311. 1997 certificate noting first year of certification. Bob Crowder collection.
90. Bob Crowder to Professor Rowe, 23 November 1992.
91. Bob Crowder, 'Under the weather', *Growing Today*, May 1993, p. 18.
92. Bob Crowder to Bernward Geier, 1 September 1992.

7: IFOAM '94

1. Bob Crowder to Right Honourable John Falloon, 28 August 1991.
2. Ministry of Agriculture and Fisheries, 'Sustainable agriculture: A policy proposal' (Ministry of Agriculture and Fisheries, Wellington, 1991), p. viii.
3. Ibid., p. 15.
4. Ibid., p. 18.
5. Ibid., p. 19.
6. Ministry of Agriculture and Fisheries, 'Policy position paper: Towards sustainable agriculture, organic farming' (Ministry of Agriculture and Fisheries, Wellington, undated), p. 20. Although undated, the reference to 1994 as a future date suggests a publication date of mid- to late 1993.
7. Bob Crowder, 'Travel Diary', 9 January 1993.
8. Bob Crowder, 'Travel Diary', 10 January 1993.
9. Bob Crowder, 'Travel Diary', 9 January 1993.
10. Bob Crowder, 'Travel Diary', 12 January 1993.
11. Bob Crowder, 'Travel Diary', 14 January 1993.
12. Bob Crowder, 'Travel Diary', 15 January 1993.
13. Bob Crowder, 'Travel Diary', 15, 16 January 1993.
14. Bob Crowder, 'Travel Diary', 17 January 1993.
15. See, for example: 'Organic work recognised', *The Press*, 5 March 1993; 'Science award', *Christchurch Star*, 6 March 1993; 'Organic husbandry work recognised', *Southland Times*, 12 March 1993; 'Organic award', *Otago Southland Farmer*, 12 March 1993; 'International award for organics pioneer', *Farm Management News*, March 1993; 'Recognition award', *Canterbury Today*, April 1993; 'International award for Lincoln organics pioneer', *Horticulture Today* (undated clipping).
16. Bob Crowder, 'Travel Diary', 18 January 1993.
17. Bob Crowder, 'Travel Diary', 21 January 1993.
18. Bob Crowder, 'Travel Diary', 22 January 1993.
19. Bob Crowder, 'Travel Diary', 23 January 1993.
20. Roland Clark, 'Amongst friends', *Growing Today*, May 1993, p. 37.
21. Ibid., p. 37. Of the apple varieties, Sunset, Tydeman's Late Orange, Discovery, Merton Worcester, Laxton's Fortune and Sturmer came out on top.

22. Bob Crowder, 'Undercover agents', *Growing Today*, May 1993, p. 39.
23. Ibid., p. 41.
24. England & Wales, National Probate Calendar (Index of Wills and Administrations), 1858–1995.
25. Bob Crowder, interview, 14 September 2021.
26. True to form, Bob, sensing Jared's interest, took him on an inspection tour for BioGro in 1991, where Jared discovered that doing that kind of work would be his 'dream job'. Jared started working for BioGro in 1999 and in 2022 was still doing this work. Jared White, interview, 6 November 2021.
27. Jared White, interview, 6 November 2021.
28. Stuart Jeffrey and Page Lawson, interview, 2 November 2021.
29. Ibid.
30. Nicole Bührs, interview, 21 October 2021.
31. Charles Merfield, interview, 26 October 2021.
32. '10th International Organic Agriculture IFOAM Conference: Preliminary programme', Bob Crowder collection.
33. Bob Crowder to Bernward Geier, 20 November 1993.
34. Bernward Geier to Bob Crowder and Don Crabb, 3 February 1994.
35. Bob Crowder, 'Travel Diary', 24 February 1994.
36. Bob Crowder to Bernward Geier, 9 March 1994.
37. See, for example, Shannon Horst (for Allan Savory) to Bob Crowder, 14 September 1994; Lorraine Harding (for Tom Harding) to Bob Crowder, 10 May 1994.
38. Brendan Hoare, interview, 4 December 2021.
39. 'Tours to supplement conference', *Acres Australia*, vol. 1, no. 12, p. 35.
40. Bob Crowder to David Lance, 25 January 1994; Don Crabb, 'Best and brightest gather for IFOAM 94', draft media release for *Soil & Health*, September 1994.
41. HRH The Prince of Wales, 'Message of Support', 29 November 1994. Bob Crowder collection.
42. Bob Crowder, 'Gathering in green', *Growing Today*, October 1994, p. 62.
43. Ibid., p. 63.
44. Minutes of the IFOAM Fayre sub-committee, 17 November 1994.
45. Bob Crowder to Padmini Krishnan (for Vandana Shiva), 4 August 1994.
46. The pōwhiri was apparently led by Maurice Gray. See Maurice Gray (Centre for Māori Studies and Research) to Don Crabb, 28 January 1993.
47. See, for example, Bob Crowder to Vicki Buck, 26 January 1995.
48. '10th International Organic Agriculture IFOAM Conference, Opening Ceremony', pamphlet, Bob Crowder collection; Bob Crowder to Padmini Krishnan (for Vandana Shiva), 4 August 1994.
49. Helen Browning, 'A journey towards organic farming – a reflection on our challenging path', paper presented at IFOAM '94. Bob Crowder collection.
50. Don Crabb, 'Best and brightest gather for IFOAM 94', draft media release for *Soil & Health*, September 1994.
51. Nor'Wester, 'Set fair for change', *Growing Today*, March 1995, p. 62.
52. Bob Crowder to Bruce Chapman, 2 August 1995.
53. Bob Crowder to Erihapeti Rehu-Murchie, 23 November 1993; Jon Manhire, interview,

21 October 2021; Bob Crowder to Morgan Williams, 20 January 1995; Bob sent letters of appreciation to all of these on 20 January 1995.
54. Biological Husbandry Unit, Publicity Income Report, 17 August 1995.
55. Bob Crowder, interview, 18 October 2021.

8: THE STRUGGLE TO SAVE THE BHU, 1995–98

1. Bob Crowder, 'Travel Diary', 11 February 1995.
2. Bob Crowder, interview, 18 October 2021.
3. Bob Crowder, 'Travel Diary', 14 February 1995.
4. Bob Crowder, 'Travel Diary', 15 February 1995.
5. Bob Crowder, 'Travel Diary', 27 February 1995.
6. Bob Crowder, interview, 27 August 2018.
7. Bob Crowder, 'Travel Diary', 1, 2, 6 and 18 March 1996.
8. Nicole Bührs, interview, 21 October 2021.
9. Bob Crowder, interview, 18 October 2021.
10. Geoff Barnett, interview, 14 September 2021.
11. Ernst rented shelf space from Brian Newbery.
12. Bruce Chapman to Bob Crowder, 2 August 1995; Luke Southorn to Bob Crowder, 11 September 1995.
13. Bob Crowder to Bruce Chapman, 2 August 1995.
14. R.E. Gaunt to B.J. Ross, 5 September 1995.
15. Ibid.
16. Ibid.
17. Bob Crowder, interview, 18 October 2021.
18. Bob Crowder to Vicki Buck, 26 January 1995.
19. Charles Merfield, interview, 26 October 2021; Charles Merfield, 'CCOG's purpose', *Canterbury Commercial Organic Group Newsletter*, no. 3, April 1998, p. 2.
20. Rod Donald to Bob Crowder, 14 January 1997.
21. Question No. 3622: Jeanette Fitzsimons to the Minister of Agriculture, received 10 March 1997. Demeter was the name of the certification standards issued by the Biodynamic Farming and Gardening Association.
22. Rod Donald to Lockwood Smith (Minister of Agriculture), 6 March 1997.
23. Bob Crowder to Lockwood Smith, 19 December 1996.
24. Lockwood Smith to Rod Donald, 27 February 1997.
25. Lockwood Smith to Rod Donald, 27 February 1997 – copy for Bob Crowder.
26. Bob Crowder to Lockwood Smith, 12 May 1997.
27. 'Bob Crowder retires', *Commercial Grower*, July 1997, p. 35.
28. Ibid.
29. Bob Crowder, 'Biological Husbandry Unit at Lincoln & the Bio-Village need our support!', *Soil & Health*, March/April 1998, p. 3.
30. Ibid.
31. Bob Crowder, 'Travel Diary', 5, 6 March 1997.
32. www.dpmc.govt.nz/publications/queens-birthday-honours-list-1997

33. Tim Chamberlain, 'Curriculum Vitae of Mr Robert Crowder', February 1996, Bob Crowder collection.
34. Peter Elworthy, quoted by Tim Chamberlain in 'Curriculum Vitae of Mr Robert Crowder', February 1996.
35. Lockwood Smith to Bob Crowder, 9 June 1997.
36. Bob Crowder to Matt Morris, 2 September 1997.
37. Ibid.
38. Bob Crowder, 'Travel Diary', 23 September 1997.
39. Bob Crowder, 'Travel Diary', 25 September 1997.
40. Ibid.
41. 'The future agenda for organic trade: 5th IFOAM International Conference on Trade in Organic Products, delegate list', September 1997.
42. Bob Crowder, 'Travel Diary', 25 September 1997.
43. Bob Crowder, 'Travel Diary', 26 September 1997.
44. Bob Crowder, 'Travel Diary', 28 September 1997. Bob recalled that Wendy was the only woman for whom he ever felt a sexual attraction; Bob Crowder, pers. comm., 28 July 2022.
45. Bob Crowder, 'Travel Diary', 29 September 1997.
46. Bob Crowder, 'Travel Diary', 30 September 1997.
47. Bob Crowder, 'Travel Diary', 1 October 1997.
48. Bob Crowder, 'Travel Diary', 2 October 1997.
49. Bob Crowder, 'Travel Diary', 4 October 1997.
50. Bob Crowder to Matt Morris, 23 October 1997.
51. Bob Crowder, 'Travel Diary', 12 October 1997.
52. Bob Crowder to Matt Morris, 2 September, 23 October, 10 December 1997.
53. 'Newsletter of the Organic Garden Organisations', Soil & Health Canterbury branch, November 1997, p. 1.
54. Geoff Barnett, interview, 14 September 2021.
55. Anne Seyger, pro forma letter, 29 August 1997, OGCT archive.
56. Minutes of the BHU 21st Committee, 28 October and 4 November 1997.
57. Ibid.
58. Bob Crowder to Vicki Buck, 4 November 1997.
59. Ibid.
60. Bob Crowder to Matt Morris, 10 December 1997.
61. Ray Wright to Trudy Burgess, 11 February 1998, OGCT archive.
62. Bob Crowder to Trudy Burgess, 23 February 1998, OGCT archive.
63. Bob Crowder, 'Biological Husbandry Unit (BHU) open day 19th April', *Soil & Health Association Canterbury Branch Newsletter*, Issue 4, May 1998, p. 2; Charles Merfield, interview, 26 October 2021.
64. Geoff Barnett, interview, 14 September 2021; Helen Duckworth interview, 19 April 2023.
65. Crowder, 'Biological Husbandry Unit (BHU) open day 19th April', p. 2.
66. Ibid.
67. Bob Crowder, 'Travel Diary', 6 July 1998.
68. Bob Crowder, 'Travel Diary', 10 July 1998.
69. Bob Crowder, 'Travel Diary', 7 July 1998.
70. Bob Crowder, 'Travel Diary', 11 July 1998.

71. Bob Crowder, 'Travel Diary', 13 July 1998.
72. Bob Crowder, 'Travel Diary', 16 July 1998.
73. Bob Crowder, 'Travel Diary', 15 July 1998.
74. Bob Crowder, 'Travel Diary', 19 July 1998.
75. Bob Crowder, 'Travel Diary', 8 August 1998.
76. Bob Crowder, 'Travel Diary', 10 August 1998.
77. Bob Crowder, 'Travel Diary', 12 August 1998.
78. Bob Crowder, 'Travel Diary', 15 August 1998.

9: THE DEATH OF THE BHU? 1998–2000

1. See, for example, 'Biological Husbandry Unit', *Organic Matters*, August 1998, p. 4.
2. Bill Kain to Joep Nederpelt, 5 August 1998.
3. Bill Kain to Trudy Burgess, 1 September 1998, OGCT archive; Trudy Burgess to Bill Kain, 7 September 1998, OGCT archive.
4. Jon Manhire, interview, 21 October 2021.
5. Brendan Hoare, interview, 4 December 2021.
6. Notes of Meeting of Organic Movement Representatives with Bill Kain and Ian Cornforth, 1 October 1998, OGCT archive.
7. Ibid.
8. Jon Manhire, interview, 21 October 2021; Jon Manhire to Matt Morris, email, 31 October 2021; 'Biological Husbandry Unit (BHU), 'The organic vision': Business plan' (Draft, April 2002), BHU Organics Trust archive.
9. Memorandum of Understanding between Lincoln University and representatives of the organic movement regarding the Biological Husbandry Unit, Lincoln University, 11 November 1998, OGCT archive.
10. Memorandum of Understanding between Lincoln University and the representatives of the organic movement regarding the Biological Husbandry Unit (BHU), Lincoln University and the expansion of teaching, demonstration and research of organic agriculture at Lincoln University', 23 April 1999, OGCT archive. A stack of versions was circulated during 1998 and early 1999, as seen in the Bob Crowder collection.
11. 'Minutes from the last S&H public meetings', *Organic Matters*, November 1998, p. 4.
12. Charles Merfield, 'BHU under threat?', *Soil & Health*, September/October 1998, p. 35.
13. Bob Crowder, 'Another look at the BHU', *Soil & Health*, November/December 1998, p. 43.
14. Anne Seyger, 'Biological Husbandry Unit', *Organic Matters*, February 1999, p. 1.
15. Ibid.
16. Shen Khen Heng, 'BHU: A student's perspective', *Organic Matters*, August 1999, p. 22.
17. Bob Crowder, 'The organic dilemma', *Organic Matters*, November 1999, p. 18.
18. Jamie Trachta-Dyson to Bob Crowder, fax, 30 March 1999.
19. Bob Crowder to Jamie Trachta-Dyson, undated (but before late May, so c. April 1999).
20. 'Cricket: Michael Watt – The saving face of cricket', *New Zealand Herald*, 23 February 2001.
21. Bob Crowder, interview, 22 December 2021.
22. Bob Crowder, 'Paradise Lost', copy emailed to Matt Morris, 24 November 1999.
23. Ivo Wynn-Williams to Michael Watt, 25 May 1999.

24. 'University plans new research centres', *The Press*, 5 May 1999.
25. Ivo Wynn-Williams to Michael Watt, 25 May 1999.
26. Michael Watt to Ivo Wynn-Williams, 25 May 1999.
27. Bob Crowder, 'Travel Diary', 2 July 1999.
28. Bob Crowder, interview, 22 December 2021.
29. Crowder, 'Paradise Lost'.
30. Frank Wood to Michael Watt, fax, 9 July 1999. Two original faxes of this letter have survived, but they are very faded and partly illegible. Fortunately, another version of the letter survives written in Bob's own hand. The only reasonable explanation for this can be that Bob was aware of the ephemeral nature of print on fax paper and wanted to preserve a copy for posterity or even simply to ensure he had a reliable back-up copy to refer to. Bob Crowder collection.
31. Bob Crowder, 'Travel Diary', 12 July 1999.
32. Bob Crowder, interview, 22 December 2021.
33. Bob Crowder, 'Travel Diary', 13 July 1999.
34. Ibid.
35. Crowder, 'Paradise Lost'.
36. Charles Merfield to Bob Crowder, Rex Verity, Holger Kahl and Brendan Hoare, 8 July 1999.
37. Letter drafted by Bob Crowder for Michael Watt to Frank Wood, undated (presumably 13 or 14 July 1999). It seems the letter was never sent.
38. Charles Merfield to Bob Crowder, Holger Kahl, Rex Verity, Brendan Hoare, 15 July 1999.
39. Ibid.
40. Brendan Hoare to Holger Kahl, Charles Merfield, Bob Crowder, Rex Verity, 15 July 1999.
41. Bill Kain to Bob Crowder, 10 August 1999.
42. Bob Crowder to Michael Watt, undated. The letter says that Kowhai Farm was launched 'last week': it was launched on 26 August 1999. Whether this letter was sent cannot be confirmed.
43. Ibid.
44. Invitation: 'Kowhai Farm: The Lincoln University Heinz Wattie's organic farm', undated, 1999. The Lincoln contact person on the invitation was Steve Wratten.
45. 'Kowhai Farm', *Organic Matters*, November 1999, p. 20.
46. 'Editorial: Going the organic way', *The Press*, 14 August 1999, p. 10.
47. Bob Crowder, 'Organic farming', Letters to the Editor, *The Press*, 25 August 1999.
48. Bob Crowder, 'The organic dilemma', *Organic Matters*, November 1999, p. 18.
49. Ibid.
50. Ibid.
51. Brendan Hoare, interview, 4 December 2021.
52. Minutes of the Inaugural Annual General Meeting of the Organic Producers Exporters Group, 12 December 1995. In due course, Jon Manhire became co-chair, then executive director.
53. Brendan Hoare, interview, 4 December 2021.
54. 'Lincoln organic workshop: Post mortem', *Canterbury Commercial Organics Group Newsletter*, no. 8, July 1999, p. 4.
55. Brendan Hoare, interview, 4 December 2021.

56. Ibid.
57. Matt Morris, 'From your trust co-ordinator', *Organic Matters,* February 2000, p. 14.
58. Brendan Hoare, interview, 4 December 2021; Mark Hill, 'S&H national meeting', *Organic Matters*, November 1998, p. 2.
59. 'BHU', *Organic Matters*, August 2000, p. 4.
60. Robyn Patchett, 'Biological Husbandry Unit Trust launched', *Canterbury Commercial Organics Group Newsletter*, no. 13, July 2000, p. 3.
61. Ibid.
62. Ibid.
63. Peter Green, 'President's message', *Organic Matters*, November 2000, pp. 2–3.
64. Bob Crowder to Peter Elworthy, 15 September 2000.
65. Ibid.
66. Peter Elworthy to Bob Crowder, 13 September 2000.
67. Matt Morris to Peter Elworthy, 6 October 2000.
68. Biological Husbandry Unit Trust, minutes of the meeting held 4 October 2000, Jon Manhire collection.
69. Bob Crowder, 'Travel Diary', 4, 5, 7 October 2000.
70. Bob Crowder, 'Reflections on the East Coast Triangle from a bicycle seat', unpublished manuscript, undated, Bob Crowder collection. See also Hugh Wilson, 'Bicycle touring: No petrol woes', *The Press*, 30 December 1974, p. 15.
71. Matt Morris, 'Editorial', *Organic Matters*, November 2000, p. 3.
72. Bob Crowder, interview, 18 October 2021.
73. Biological Husbandry Unit Trust, minutes of the meeting held 4 October 2000, Jon Manhire collection.

10: RETIREMENT, 2001–23

1. Biological Husbandry Unit Trust, Minutes of the meeting held 10 August 2001, Jon Manhire collection; Application for Incorporation of Trustees as a Board for the Biological Husbandry Unit Organics Trust, 12 June 2002: https://register.charities.govt.nz/CharitiesRegister/ViewCharity?accountId=0fe814ae-0bb5-dd11-9e5d-0015c5f3da29&searchId=95f77d87-812a-4f5a-b9b8-5b2d443f72e5
2. Mollie Chalklen served on the board until mid-2005. She resigned for health reasons. In early March 2006 Jon Manhire reported that she had suffered a stroke and was in hospital. BHU Organics Trust Board minutes of the meeting 19 August 2005; 3 March 2006, BHU Organics Trust archive.
3. Jon Manhire, interview, 21 October 2021.
4. Tim Jenkins to Matt Morris, email, 14 November 2021.
5. Tim Jenkins to Matt Morris, email, 14 November 2021.
6. Tim Jenkins to Matt Morris, email, 14 November 2021.
7. Sir Peter Elworthy, quoted in Mark Hill, 'First BHU open day of 2002', *Organic Matters*, May 2002, p. 4.
8. Sir Peter Elworthy, quoted in Mark Hill, 'First BHU open day of 2002', *Organic Matters*, May 2002, p. 5.

9. Mollie Chalklen, quoted in Mark Hill, 'First BHU open day of 2002', *Organic Matters*, May 2002, p. 5.
10. Lease Agreement between Lincoln University and the Biological Husbandry Unit Organics Trust, signed 11 April 2002 (by Lincoln University) and 15 April 2002 by the BHU Organics Trust, Jon Manhire collection.
11. Peter Elworthy to Companies Office, 1 July 2002, BHU Organics Trust archive. The deed had been signed by all parties on 12 June 2002.
12. Biological Husbandry Unit Organics Trust, Minutes of the meeting held 12 December 2001, BHU Organics Trust archive.
13. Tim Jenkins, quoted in Mark Hill, 'Birthday celebrations at the Biological Husbandry Unit', *Organic Matters*, Summer 2002, p. 13.
14. Tim Jenkins to Matt Morris, email, 14 November 2021.
15. Tim Jenkins, 'Alf Annual Report Biological Husbandry Unit: March 2003 to August 2003', p. 2, BHU Organics Trust archive.
16. Tim Jenkins to Matt Morris, email, 14 November 2021.
17. Charles had commenced a postgraduate diploma originally but needed to halt work on this due to the death of his stepfather in Britain. On his return he commenced the master's. Charles Merfield to author, 24 August 2022.
18. Charles Merfield, interview, 26 October 2021. His PhD was completed in 2006: 'Organic F1 hybrid carrot seed (*Daucus carota* L.) production : The effect of crop density on seed yield and quality, thermal weeding and fungal pathogen management' (Lincoln University, 2006); Charles Merfield to author, 24 August 2022.
19. Bob Crowder, interview, 18 October 2021.
20. Matt Morris, quoted in 'Organic Garden City Trust to Administer local organic certification scheme', *Canterbury Organics Group Newsletter*, no. 14, February 2001, p. 5. By October 2001 a draft constitution for a new organisation to manage the scheme locally had been drawn up. The new organisation, Canterbury Organic, was incorporated in November 2001. See Robyn Patchett, 'Progress with Canterbury Organics', in *Canterbury Organics Group Newsletter*, no. 17, October 2001, p. 3, and 'Canterbury Organic', in *Canterbury Organics Group Newsletter*, no. 20, July 2002, p. 2.
21. Matt Morris, 'Organic Farm NZ – Launched', in *Organic Matters*, Summer, 2002.
22. Matt Morris, 'Canterbury Organic Update', in *Canterbury Organics Group Newsletter*, no. 21, November 2002, p. 2.
23. See, for example, Bob Crowder, 'Travel Diary', 28 February 2003.
24. Bob Crowder, 'Travel Diary', 31 March 2004.
25. Bob Crowder, interview, 18 October 2021.
26. Ibid.
27. Ibid.
28. Bob Crowder, 'Travel Diary', 20 June 2004.
29. Bob Crowder, 'Travel Diary', 23 and 24 June 2004.
30. Bob Crowder, 'Travel Diary', 30 June and 1 July 2004.
31. Bob Crowder, 'Travel Diary', 9 December 2007.
32. 'Workshops for organic growers at BHU', *Canterbury Commercial Organics Group Newsletter*, no. 20, July 2002, p. 3.
33. Tim Jenkins to Matt Morris, email, 14 November 2021.

34. Tim Jenkins, 'BHU update', *Organic Matters*, Autumn 2003, p. 9.
35. Tim Jenkins to Matt Morris, email, 14 November 2021.
36. Jon Manhire, interview, 21 October 2021.
37. Tim Jenkins to Matt Morris, email, 14 November 2021. An afternoon tea was held on 7 December 2004 to thank Tim for his time at the BHU. However, concern was expressed even as late as August 2005 that Tim was still responsible for meeting several BHU obligations, but that he was doing all of this voluntarily. BHU Organics Trust Board Minutes of the Meeting 7 December 2004, 19 August 2005. BHU Organics Trust archive.
38. Tim Jenkins to Matt Morris, email, 14 November 2021.
39. Ibid.
40. BHU Organics Trust Management Committee Minutes of the Meeting 10 March 2005, 12 April 2005. BHU Organics Trust archives.
41. 'NZ Organic Food and Wine Festival', *Canterbury Commercial Organics Group Newsletter*, no. 14, February 2001, pp. 2, 8.
42. John Coburn, 'Bob Crowder on organic principles, myths and magic', *Organic Matters*, Autumn 2003, p. 5.
43. Bob Crowder, 'The art of being a good grower', *New Zealand Lifestyle Block*, April 2014, pp. 72–73.
44. Klaus Thoma, interview, 8 January 2021. Golden Bay Community Gardens was founded by Sue Shotton, who was part of the Nelson Morris side.
45. Sol Morgan, interview, 10 March 2020. Bob was patron until 2007. Sol Morgan to Matt Morris, pers. comm., 11 December 2021.
46. Bob Crowder, 'Travel Diary', 1, 2 and 5 August 2003.
47. Bob Crowder, 'Travel Diary', 18, 19, 20 and 21 September 2003.
48. Bob Crowder, 'Travel Diary', 1 July 2004. The two continued to make contact at Christmas as late as 2022; Bob Crowder, pers. comm., 28 July 2022.
49. Klaus Thoma, interview, 8 January 2021.
50. Bob Crowder, 'Travel Diary', 4 January 2012.
51. Brendan Hoare, interview, 4 December 2021.
52. Ibid.
53. Terry Parminter, Neels Botha and Jon Tanner, 'A proposed organisational structure for Organics Aotearoa New Zealand to provide an extension service to farmers and growers', *Extension Farming Systems Journal*, vol. 2, no. 1: www.researchgate.net/publication/264852834_A_proposed_organisational_structure_for_Organics_Aotearoa_New_Zealand_to_provide_an_extension_service_to_farmers_and_growers_Growth_in_organic_production_in_New_Zealand/link/54d1cac40cf28959aa7b86aa/download; Brendan Hoare, interview, 4 December 2021.
54. BHU Organics Trust Board, Minutes of the Meeting 3 March 2006, BHU Organics Trust archive.
55. Ibid.; Holger Kahl to BHU Board of Trustees, 10 February 2007, BHU Organics Trust archive.
56. Holger Kahl to Matt Morris, 3 February 2022.
57. Jon Manhire, interview, 21 October 2021; Holger Kahl to Matt Morris, 3 February 2022.
58. 'Organic advisory programme ends', 30 June 2009: www.scoop.co.nz/stories/BU0906/S00692/organic-advisory-programme-ends.htm

59. BHU Organics Trust Board, Minutes of the Meeting 2 June 2006, BHU Organics Trust archive.
60. Holger Kahl, 'Manager's report to the BHU Board of Trustees', 20 October 2006; Roger Field to Holger Kahl, 9 October 2006, BHU Organics Trust archive; BHU Organics Trust Board, AGM Minutes, 20 November 2012, BHU Organics Trust archive.
61. Bob Crowder, 'Travel Diary', 19 July 2008.
62. Ibid.
63. Bob Crowder, 'Travel Diary', 21 November 2008.
64. Bob Crowder, 'Travel Diary', 23 November 2008.
65. Bob Crowder, interview, 18 October 2021.
66. Such programmes had developed through the University of California, partly in response to Hatch Act violations in the 1960s, as discussed in Chapter 3.
67. Bill Martin to Matt Morris, email, 19 October 2021.
68. Bill Martin, interview, 18 October 2021.
69. Bob Crowder, interview, 18 October 2021.
70. Jeanette Fitzsimons, quoted in 'Welcome to our patrons', *Organic NZ*, November/December 2010, p. 56.
71. Bob Crowder, 'Travel Diary', 25 September 2010.
72. Bob Crowder, 'Travel Diary', 22 August 2011. Those present will recall the very heated debate between Bob and Steffan Browning on the topic of a proposed merger of Soil & Health with BioGro.
73. Bob Crowder, interview, 18 October 2021; Soil & Health Association of New Zealand, Minutes for the AGM, 31 May 2014.
74. Bob Crowder, 'Catastrophe and the way forward', *Organic NZ*, November/December 2010, p. 22.
75. Ibid., p. 23.
76. Bob Crowder, interview, 18 October 2021.
77. Bob Crowder, 'Travel Diary', 4–7 March 2011.
78. Bob Crowder, interview, 18 October 2021. Bob later believed this was similar to long Covid. Bob Crowder, pers. comm., 28 July 2022.
79. Bob Crowder, interview, 18 October 2021.
80. Bob Crowder, 'Travel Diary', 25 and 26 May 2013.
81. Bob Crowder, 'Travel Diary', 23, 29 May; 3, 13 June; 1 July 2013.
82. Bob Crowder, 'Travel Diary', 2, 3, 5 July 2013.
83. Bob Crowder, 'Travel Diary', 10, 19 July 2013.
84. 'Fruit and Nut Tree Hui – Feeding Our Futures', Minutes, 21 September 2013, FRN archive.
85. Food Forest Steering Group (Feeding our Futures), Minutes, 30 November 2013, FRN archive.
86. For example, Matt Morris to Steering Group, email, 3 March 2014. The formal decision to call it that was made in June 2014. Food Resilience Network minutes, 25 June 2014, FRN archive.
87. Bob Crowder to Matt Morris, email, 30 July 2014.
88. Bob Crowder to Tony Moore, email, 24 August 2014.
89. Peter Wells to Matt Morris, email, 23 December 2021.

90. Charles Merfield, interview, 26 October 2021.
91. Ibid.; Charles Merfield to author, 24 August 2022.
92. BHU Organics Trust Board, Minutes of the meeting 7 November 2005, BHU Organics Trust archive.
93. Ibid.
94. Charles Merfield, interview, 26 October 2021.
95. Ibid.
96. Ibid.
97. 'BHU review and strategy update', October 2021, BHU Organics Trust archive.
98. Jon Manhire, interview, 21 October 2021.
99. 'Lease Agreement', signed by Bruce McKenzie, 18 November 2021 and Jon Manhire, 19 November 2021.
100. 'Time for action: 2020/21 New Zealand organic sector market report', Organics Aotearoa New Zealand, 2021, p. 4.
101. Bob Crowder, interview, 18 October 2021.
102. Madge Crowder to Bob Crowder, 29 October 1983.
103. Charles Manhire, interview, 26 October 2021.
104. Bob Crowder, quoted in Mark Hill, 'Lincoln University workshop,' *Canterbury Commercial Organics Group Newsletter*, no. 8, July 1999, p. 3.
105. See Lawrence Woodward, 'Health, sustainability and the global economy: The organic dilemma', Elm Farm Research Centre, September 1997.
106. Bob Crowder, interview, 18 October 2021.
107. Ibid.
108. Bob Crowder, Lecture, 'The history of organics and the BHU', Organic Training College, BHU, 14 February 2022.
109. Bob Crowder, 'Education in organics: Practising that which is preached' (undated, but likely 1996. Probably presented at IFOAM), p. 10. Although in this draft he incorrectly attributed the idea to Voltaire.
110. Brendan Hoare, interview, 4 December 2021.

BIBLIOGRAPHY

BOOKS AND JOURNALS

Balfour, Eve, *The Living Soil and the Haughley Experiment* (London: Faber & Faber, 1943)

Barton, Gregory, *The Global History of Organic Farming* (Oxford: Oxford University Press, 2018)

Carlisle-Cummins, Ildi, 'From ketchup to California cuisine: How mechanical tomato harvesting prompted today's food movement' (2015): https://news.plantsciences.ucdavis.edu/2015/07/24/how-the-mechanical-tomato-harvester-prompted-the-food-movement/

Crowder, R., 'Mechanical harvesting of tomatoes', *New Zealand Journal of Agriculture*, vol. 129, no. 5, November 1974, pp. 40–46

Crowder, R., 'Onion production: A comparison between yield and grading', *New Zealand Journal of Agriculture*, vol. 127, no. 5, November 1973, pp. 26–27

Crowder, R., 'Pelleted seed in onion production', *New Zealand Journal of Agriculture*, vol. 121, no. 5, November 1970, pp. 80–85

Crowder, R., 'Research into direct seeding of asparagus', *New Zealand Journal of Agriculture*, vol. 125, no. 1, July 1972, pp. 43–45

Crowder, R.A., 'New techniques could boost vegetable growing', *New Zealand Journal of Agriculture*, vol. 118, no. 3, March 1969, pp. 78–81

Crowder, R.A., 'Vegetable with a difference: The witloof, widely grown in Europe is being grown in Pukekohe', *New Zealand Journal of Agriculture*, vol. 111, no. 7, December 1965, pp. 35

Crowder, R.A., Development of Machine Methods for Extensive Crop Planting: Onions (Lincoln College/University of Canterbury, 1968)

Crowder, R.A., *Extensive Vegetable Production* (Lincoln College/University of Canterbury, 1968)

Crowder, R.A., *Preliminary Investigations into Large-Scale Intensive Production of Tomato* (Lincoln College/University of Canterbury, 1968)

Dawson, Bee, *A History of Gardening in New Zealand* (Auckland: Godwit, 2010)

Dollimore, Jonathan, *Sexual Dissidence: Augustine to Wilde, Freud to Foucault* (New York: Oxford, 1991)

'Drama', *Lincoln College Magazine*, no. 97, 1972: https://livingheritage.lincoln.ac.nz/nodes/view/604#idx3629

Freud, Sigmund, '"Civilized" sexual morality', *Civilization, Society and Religion*, vol. 12 (London: Penguin, 1985)

Hayward Lorna (ed.), *On the Crest of the Hill: Devizes Grammar School, 1906–1969* (Salisbury: Hobnob Press, 2006)

Hesse, Hermann, *Steppenwolf* (London: Penguin, 2009)

Hoffelmeyer, M., '"Out" on the farm: Queer farmers manoeuvring heterosexism and visibility', *Rural Sociology*, vol. 86, no. 4, 2021, pp. 752–76

Howard, Albert, *An Agricultural Testament* (Oxford: Oxford University Press, 1940)

Hudson, R. & Calvert, J. (eds), *Ecological Agriculture: Review and annotated bibliography* (Wellington: Federated Farmers of New Zealand Inc., 1983)

Jarman, Derek, *At Your Own Risk: A saint's testament* (Vintage: London, 1992)

Leslie, I.S., 'Queer farmers: Sexuality and the transition to sustainable agriculture', *Rural Sociology*, vol. 82, no. 4, 2017, pp. 747–71

Lockeretz, W. (ed.), *Organic Farming: An international history* (Oxfordshire: CAB International, 2007)

McVarish, Scott, *The Greening of New Zealand: New Zealanders' visions of green alternatives* (Auckland: Random Century, 1992)

Merfield, Charles, 'Organic F1 Hybrid Carrot Seed (*Daucus carota L.*) Production: The effect of crop density on seed yield and quality, thermal weeding and fungal pathogen management' (Lincoln University, 2006)

Ministry of Agriculture and Fisheries, 'Sustainable agriculture: A policy proposal' (Ministry of Agriculture and Fisheries, Wellington, 1991)

Ministry of Agriculture and Fisheries, 'A proposed policy on organic agriculture' (Ministry of Agriculture and Fisheries, Wellington, 1991)

Ministry of Agriculture and Fisheries, 'Policy position paper: Towards sustainable agriculture, organic farming' (Ministry of Agriculture and Fisheries, Wellington, undated)

Morris, Matt, *Common Ground: Garden histories of Aotearoa* (Dunedin: Otago University Press, 2020)

Morrison, T.M., 'University raises status of study in horticulture', *New Zealand Journal of Agriculture*, vol. 114, no.1, January 1967, pp. 58–61

Oswin, N., 'Critical geographies and the uses of sexuality: Deconstructing queer space', *Progress in Human Geography*, vol. 32, no. 1, 2008, pp. 89–103

Parminter, Terry, Botha, Neels & Tanner, Jon, 'A proposed organisational structure for Organics Aotearoa New Zealand to provide an extension service to farmers and growers', *Extension Farming Systems Journal*, vol. 2, no. 1: www.researchgate.net/publication/264852834

Paul, John, 'The lost history of organic farming in Australia', *Journal of Organic Systems*, vol. 3, no. 2 (2008)

Roche, M., 'An Interventionist State: "Wise-use" forestry and soil conservation', in E. Pawson & T. Brooking (eds), *Making a New Land: Environmental histories of New Zealand* (Dunedin: Otago University Press, 2013)

Ross, Kirstie, *Going Bush: New Zealanders and nature in the twentieth century* (Auckland: Auckland University Press, 2008)

Shearer, Georgina, et al., 'Crop production costs and returns on Midwestern organic farms: 1977 and 1978', *American Journal of Agricultural Economics*, vol. 63, no. 2, 1981, pp. 264–96

'Time for action: 2020/21 New Zealand organic sector market report' (Organics Aotearoa New Zealand, 2021)

Woodward, Lawrence, 'Health, sustainability and the global economy: The organic dilemma', (Elm Farm Research Centre, 1997)

INTERVIEWS

Geoff Barnett, 14 September 2021
Jo Blakely, 18 October 2021
Nicole Bührs, 21 October 2021

Bob Crowder, 27 August 2018, 10 September 2018, 18 September 2018, 25 September 2018, 1 October 2018, 9 October 2018, 20 November 2018, 8 January 2019, 25 March 2019, 21 May 2019, 14 January 2020, 10 March 2020, 2 June 2020, 14 September 2021, 18 October 2021, 22 December 2021, 7 February 2022

David Dennis, 23 April 2019

Bob Douglas, 23 July 2019

Helen Duckworth, 19 April 2023

Brendan Hoare, 4 June 2020, 4 December 2021

Stuart Jeffrey and Page Lawson, 2 November 2021

Jon Manhire, 21 October 2021

Tim Maples, 3 March 2020

Bill Martin, 18 October 2021

Chris and Jenny May, 19 May 2020

Charles Merfield, 26 October 2021

Sol Morgan, 10 March 2020, 11 December 2021

David Musgrave, 3 January 2022

Perry Spiller, 9 June 2020

Klaus Thoma, 8 January 2021

Peter Waugh, 22 June 2020

Jared White, 6 November 2021

COLLECTIONS

Biological Husbandry Unit collection

Biological Producers and Consumers Society collection

Bob Crowder collection

Food Resilience Network collection

Soil & Health Association of New Zealand collection

NEWSPAPERS AND PERIODICALS

Canterbury Commercial Organic Group Newsletter
Central Otago News
Christchurch Star
Commercial Grower
The Dominion
Farm Management News
Growing Today
Horticulture Today
New Zealand Herald
New Zealand Lifestyle Block
Organic Garden City Trust Commercial Group Newsletter
Organic Garden Organisation Newsletter
Organic Growing
Organic Matters
Organic NZ

Otago Daily Times
Otago Southland Farmer
The Press
Soil & Health
Soil & Health Association, Canterbury Branch Newsletter
Southland Times
Straight Furrow
The Wiltshire Gazette

ACRONYMS

BHU Biological Husbandry Unit
BT *Bacillus thuringiensis*
CCOG Canterbury Commercial Organic Group
DSIR Department of Scientific and Industrial Research
ECO Environmental Conservation Organisation
GATT the General Agreement on Tariffs and Trades
IFOAM International Federation of Organic Agricultural Movements
LEO Lincoln Environment Organisation
MAF Ministry of Agriculture and Fisheries
NASAA National Association for Sustainable Agriculture – Australia
NZBPC New Zealand Biological Producers Council
NZBPCC New Zealand Biological Producers and Consumers Council
OANZ Organics Aotearoa New Zealand
OFNZ Organic Farm NZ
OFoNZ Organic Federation of New Zealand
OGCT Organic Garden City Trust
OPENZ Organic Products Exporters of New Zealand
PAN Pesticides Action Network
Project GRO Giving to Research in Organics
SSPOPs Small Scale Producers Organic Programme
TRADENZ Trade and Enterprise New Zealand
USDA US Department of Agriculture
WWOOF Weekend (later 'Willing') Workers on Organic Farms

INDEX

Page numbers in **bold** refer to images.

accreditation *see* certification
Adam, Robert 189
agricultural and horticultural chemicals 12, 13, 51, 52, 81, 85, 101, 107, 113, 117, 137, 195; *see also* herbicides; pesticides
Agropolis community garden 194, 195
Alpine Sports Club, Auckland 39–40
Altieri, Miguel 87, 103, 117, 135, 145
amenity horticulture 75
American Peace Corps 64
Asilomar State Marine Reserve organic conference, 1993 135–36
asparagus trials 60, **61**
Attic Coffee House, Christchurch 95
Australia Party 59
Austrian Alps 28, 31
Autumn Farm, Tākaka 187, 188

Balfour, Eve 180
The Living Soil 76
Banks Peninsula 49
Banks Peninsula Track 177
Barnett, Geoff 85, **102**, 103, 114, 115, 128, 140, 157, 158, 162–63, 172, **178**, 180, 181–82, 189, **xv**
Barton, Gregory 9, 80
Bath 67–68, 72
Bath City Morris **70**, 71, 72
Bath Community Design Workshop (later Comtek) 71
Baylotan insecticide 45
Beauchamp, Sue 128, 140, 148, 157
Belben, Mrs (neighbour, Isle of Wight) 19, 40
Bergland Report (1980) 81, 84
BHU Organic Training College 188–90, 195, 198–99
BHU Organics Trust 162, 172, 173–74, 177, 179–81, 186
biodiversity 115, 137, 159, 185, 196, 198
Biodynamic Farming and Gardening Association 88, 106, 108, 118, 119, 125
biodynamics 76–77, 85
Bio-Fach International Organic Trade Fair, 1993 134–35, 147
BioGro certification programme 104, 107–09, 112, 122, 131, 170, 182, 184
 25th anniversary 189
 Bob's role in development 98, 101, 126, 127, 129, 154, 182, 190–91
 professionalisation 106–07, 117, 126
 relations with IFOAM 114, 117, 118, 120–21
 relations with MAF 108, 113–14, 124, 125, 133, 150
 relations with NASAA 119, 121
biological husbandry 84, 85, 86, 87, 88, 90, 91, 96–97, 179
 aims 92
 see also organic agriculture and horticulture; and under Lincoln College
Biological Husbandry Unit *see* Lincoln College: Biological Husbandry Unit (BHU)
Blakely, Joanne, 'Organic Horticulture' 101
Borel, Jerome **178**
Boston 72
Box, Wilf 24
Bracero Program 53
British Organic Conference, 1993 134
British Soil Association 104
 Walnut Farm and Elm Farm 84
Browning, Helen 143
Browning, Steffan 188
Buchanan, Keith 57
Buck, Vicki 126, 143, 149–50, 157–58, 199
Bührs, Nicole 140, **141**, 148, 157, 158, 172
Burma campaign, World War II 17
Burwell, Karma **183**
Bussell, Kevin 63, 64

C. Alma Baker Trust 98, 127
California Agrarian Action Project 53–54
Californian Certified Organic Foods group 109
Calvert, John 90
Canterbury A&P Show, 1983, Soil Association exhibition 92–93, **93**, 197
Canterbury Commercial Organics Group (CCOG) 150, 163, 172
Canterbury Herb Society 157
Canterbury Organic 184
Canterbury Organic Producers (COP) group 100, 107, 109, 150

Canterbury Permaculture Group 140
Canterbury University *see* University of Canterbury
Canterbury University Academic Mobilisation 57
Carlisle-Cummins, Ildi 53
carrot seed, organic 181
Cephaelis ipecacuanha 28
certification
   Demeter standard 150, 170
   IFOAM international accreditation programme 107, 111, 117, 120–21, 122, 123, 124, 125, 127, 152
   non-exporting growers and manufacturers 182–84
   see also BioGro certification programme
certified organics 152
Chalklen, Mollie 80, 85, 86, 87, 89, 90–91, 174, 177, 179, 180
Chamberlain, Brian 111, 112
Chamberlain, Tim 100, 140, 152–54, **153**, 172, 181
Chapman, Guy 180
Charles, David 26, **29**, **30**
Charles, Prince of Wales (now King Charles III) 12, 142, 143
chemicals *see* agricultural and horticultural chemicals; fertilisers; pesticides
Chindit operations, World War II 17
Chiverton family, Isle of Wight 19, 21, 24
Christchurch City Council 193, 194
Christchurch Community Gardens Association 150
Christchurch Food Forest Collective 193
Christchurch Polytechnic Institute of Technology (CPIT) 157, 158, 172, 186, 188
Clack, Jean 25
Clark, Roland 136–37, 145
Cohen, Ben 135
Coleman, Bruce 43–44
Collins, Bruce 107
Commercial Grower 130, 151–52
Commonwealth Games, 1974 67
community gardens 183, 194
Community & Home Gardens Group 150
composting 76, **77**
   Biological Husbandry Unit **83**, 85, 122, **iii**
   Indore-style heaps 76
Cornforth, Ian 124, 161
Crabb, Don 141, 144

Craigieburn ski field 49
Craik, Stephen 65, **155**, 192
crop rotation *see* rotation
Crowder, Albert 16
Crowder, Bert 21, 62
Crowder, Bill **14**, 15, 17, 18, 21, 24, **27**, 65, 67, 68, 74
   Bob's visits 55, 71
   death, and scattering of ashes in Devizes 138, 147
   emigration to New Zealand 127–28
   letters to Bob 33, 37, 40, 42, 59, 62, 81
   visits to New Zealand **94**, 95, 127, **v**
Crowder, Bob
   awards and recognition
   IFOAM Recognition Award, 1993 135, 152
   Member of the Order of Merit, 1997 152–54, **153**, 197
   birth and early life 14, 15–19, **20**, 21
   impact of World War II bombing 15–16
   play with local children 19
   education
   Lincoln College PhD 46–47, 52, 59–60, 62, 66, 67, 72, 74–75
   Nottingham University 25–26, 28, **29**, **30**, 31
   primary schooling 17, 21
   secondary schooling 21, 22–23, 25, **27**
   health
   arthritis 199
   back pain 19, 26, 31, 34, 39, 40, 58, 199
   myalgic encephalomyelitis (ME), 1982 and 2011 88, 192
   homes
   73 Ashgrove Terrace, Christchurch 81, **82**, 84, **178**, 181–82, 184, 199, **i**, **iv**, **v**, **vi**, **xii**, **xiv**, **xv**
   apartment, Christchurch 199
   Auckland flats 36, 37, 39
   Bath flat 71
   Cashmere Road, Christchurch 81, **82**
   Colombo Street, Christchurch 81
   Devizes **20**, 24
   Isle of Wight 15–16
   Miles Warren flats, 64 Carlton Mill Road, Christchurch 62, 63, **63**, 68, 72, 74, 81
   Pukekohe **45**
   Wairakei Road flat, Auckland 37, 39
   homosexuality
   after arrival in Auckland, 1963 36, 37

in Bath 71
Bob's views on homosexuality 34
in Christchurch 62, 64, 65–66, 81, 100, 129
early feelings 21–22, 24–25
encounter in Panama 35
relationship with Chris Weeks 71, 74, 81
relationship with David Dennis 39–40
relationship with Murray Scott 95, **146**, 147, 156
schoolmates 22
sublimation 28, 31
legacy 154, 180, 182, 196–99
name
change from Robert to Bob 28
nickname 'Gascoin' 28
personal interests (*see also* gardening (in main index))
climbing and outdoor activities 12, 28, 31, **32**, **38**, 40, **41**, 44, 47, 49, **51**, 58, 177, **i**
folk music 71
hitchhiking trips 25, 26, 28, 40
Morris dancing 12–13, **70**, 71, 72–74, **73**, 81, 100, 121, **144**, 156, 174
reading 21
rowing 25, 26, 39
sport 21, 22, 24, 25, 28, 39–40
theatre 21, 26, **27**, 60
weather 18, 22–24, 39, 49, 131
personal travel
Britain, 1995 147
Britain, 1997 154, **155**, 156
Britain, 1999 164
Britain, 2001 184, **185**
Britain, 2011 and 2012 192
Britain and Germany, 2004 184–85
Britain and Germany, 2013 192–93
Fiji and Tonga, 1972 64–65
Perth, 1995 147
Poland, Germany and the UK, 1998 158–59
political interests 12, 44–45, 54, 57–58, 60
religion and church activities 12, 13, 42–43, 46, 58–59
social life
Auckland 37, 40, 42
Christchurch 50, 95
dancing 21, 24, 25, 26
Lincoln College 50, 60
parties, 64 Carlton Mill Road, Christchurch 62, **63**

relationship with Wendy 28, **29**, 37, 156, 159
a social person 12
youth group, Pukekohe **45**, 45–46
working life (*see also* Lincoln College (in main index))
apple and pear pruning 44
BHU Organic Training College 190
Department of Agriculture 31, 36, 43–44, 45, 46
international travel, 1993 134–36
journey to New Zealand 33–36
refresher leave from Lincoln, 1975 67–69
self-funded study tour, 1981 84–85
study tour of North America and Europe, 1969 52, 53–55
study tour of United States, 1973 66
University of Bath, School of Biological Sciences 67–68, 71
working life in organics
BioGro certification 108, 126, 127, 129, 133
IFOAM 103–04, 109, **110**, 111, 119, 120, 124–25, 126–27, 129–30, 131, 134–35, 141–45, 147, 149, **151**, 152–54, 156, 158–59
interaction with organic sector after retirement 181–82, 184, 186–87, 190, 193–95
NASAA 121
New Zealand Biological Producers Council (NZBPC) 105, 107, 108–09, 113–14, 117, 118, 121, 123, 124–25, 126–27
Organic Growers Council of New Zealand 88
Soil & Health Association 80, 90, 91, 92–93, 97, 115, 122, 162, 190–91
writing
articles 46, 52–53, 55, 60, 66, 67, 91, 187
'Extensive Vegetable Production' (1970) 55
journal 23, 33, 34, 35, 36, 37, 42, 49
monthly reports on meteorology to *Wiltshire Gazette* 23–24, 25
Crowder, Christel 159
Crowder, Cyril 12, 53
Crowder, David **14**, 15, 16, 17, 19, **20**, 21, 24, **27**, 28, 37, 58, 159, 184–85, **185**
Crowder, Elizabeth 16, 21, 40
Crowder, Hugo 192, 193
Crowder, Jack 31, 62

Crowder, Madge 14, 15–16, 17, 18, 21, 24, **27**, 28, 42, 65
  Bob's visits 55, 71
  death and scattering of ashes 184, **185**
  emigration to New Zealand 127–28, **xv**
  letters to Bob 33, 37, 39, 40, 58–59, 60, 62, 66, 67, 69, 74, 197
  in retirement home, Christchurch 138, 147, 156
  visits to New Zealand **94**, 95, 127, **v**
Crowder, Percy 62
Crowder, Simon 58, 159, 185, **185**, 192, 193
Crowder's Delight globe artichoke 7, 63, 181
Cultivate Christchurch 194–95
Czechoslovakia 55

Daly, Mike 114, 128, 149
Davis, Mrs (neighbour, Isle of Wight) 15, 17, 18
de Waal, Peter and Margaret 44–45, 57
Demeter standard 150, 170
Dennis, David 36, 37, 39–40
Department of Agriculture, New Zealand 31, 36, **43**, 43–44, 45, 46
  advisory and practical work 44
Department of Scientific and Industrial Research (DSIR) 89, 96
Devizes Grammar School 21–23, 25, **27**
Devizes, Wiltshire 21–24, 37, 128, 134, 147, 154, 156, 192
Dickinson, John 59, 65
Dieldrin 51
Donaghy, Rose 100, 128, **153**
Donald, Rod 148, 150–51, 182
Doubleday Research Association 88, 106
Douglas, Bob 51–52, 55, 59, 66–67, 72, 98
Duckworth, Helen **141**, 158, 172

earthquakes, 2010 and 2011 191–92, 193
Ecological Agriculture: Review and annotated bibliography (Hudson and Calvert, 1983) 89–90
ecological farming 75
ecosystem protection 80
Ecosystem Services 181
effective microorganisms (EM) 180
Ellyard, Peter 143
Elm Farm Research Centre, Britain 87, 192
Elworthy, Lady Fiona 145, 174, 179, 186

Elworthy, Sir Peter 87, 143, 145, 153, 173, 174, 179–80, 186
Environmental Conservation Organisation (ECO) 105
Enviroschools programme 149
Erewhon Morris 72–74, **73**
Erewhon Station 73–74
European Economic Community 134
exports 12, 111, 112, 122, 133–34, 135, 152, 161, 162, 170, 196, 198
  certified kiwifruit 108

family farming 10, 54, 104
Federated Farmers 84, 86–87, 88, 90, 91, 96, 111, 190
  commissioning of literature review 89
  request to Bob for ideas on research priorities 88–89
  Rural Land Use Committee 86–87
fertilisers 13, 50, 53, 60, 87, 92, 96, 108, 111, 185, 197
Fiji 36
Fischer, Mic 187
Fishlock, Mr (neighbour, Devizes) 24, 25
Fitzsimons, Jeanette 150, 172, 182, **183**, 184, 190, 191
flame weeding 128
Flock House *see* Seminar on Biological Farm Management Techniques (Federated Farmers, 1982)
food
  biological husbandry aims 90, 92
  genetically modified food 136, 168
  growing food in Christchurch city 183–84, 193–95
  MAF support for organically grown food 111, 112, 122
  market for organic food 92, 130, 148, 173, 196, 198
  produced by BHU 7, 85, 157, 170
  regulations and standards 126, 133
food forests 187, 189, 193
Food Resilience Network 194, 195
Frei, Ernst 100, 148, 172, 199
Freud, Sigmund 22
Fritz, Sandy 117, 118, 119
Future Farming Centre 195, 196

gardening
  73 Ashgrove Terrace, Christchurch **11**, 81,

82, 84, 181, 187, 192, 195, **i**, **iv**, **v**, **vi**, **xii**, **xiv**, **xv**
  Devizes, Wiltshire 20, 22, 24, 37, 39, 42, 71
  Isle of Wight 18–19
GATT (General Agreement on Tariffs and Trade) 143
Geering, Lloyd 145
Geier, Bernward 104, 122, 128, 134, 156, 185
  IFOAM conference New Zealand, 1994 141–42, 143
  IFOAM secretary-general 109, 117, 118, 120, 123, 124, 127, 133, **151**, 159, 197
  New Zealand visit, 1988 118, 119
genetic engineering 135–36, 168, 172, 190
gherkin research 56, 66, 84
Gill, Howard 59–60, 66
Gill, Titi 59
Gillingham, Allan 145
Gips, Terry 104
Gitananda, Swami 76
globe artichokes 63
Goks, Chinese market gardeners 44
Goldcrum Hayward 56, 66, 72
Golden Bay 184, 187–88
Good Gardeners 162–63, 187
government policy 105–06
grape industry 56
Green House, Christchurch 95, 147
Green movement, New Zealand 74
Green Party 150, 170, 172, 182, 188, 190
  Eco-Nation policy 170
Green, Peter 173
Griffith, Katherine 136
Growing Today 136, 137, 145, 187
Guyton, Robert 189

Habicht, Tyrol 28
Hamlin, Jill **183**
Hancock, Graham 24, 25
Hardie Boys, Sir Michael **153**, 154
Harding, Tom 143
Hat and Feather folk music club, Bath 71
Hatch Act 1887 (US) 53–54
Haystead, Anthony 112
Hayward, Mr (Bob's teacher) 17
Hazeldine, Mark 145
Health Foods Standards Committee 126
Henderson, Ian 100, 174, 177
Heppenheim, Germany 134
herbicides 51, 52, 56, 72, 136

Higginson, Terry **183**
Hill, Stuart 145, 170
Hillary, Sir Edmund 47
Hoare, Brendan 8, 100, **113**, 126, 131, 167, **167**, 170, 172, 180, **183**, 184, 186, 188, 199
  Biological Husbandry Unit 115, 122, 128, 161, 162, 170, 199
  Soil & Health Association president 158, 161, 172, 182
Holyoake, Keith 60
homosexuality 34
  Lord Edward Montagu trial, 1954 22
  see also under Crowder, Bob
Horticultural Market Research Unit 101
horticultural therapy 18
horticulture 50, 55
  hand-picking 59, 66
  mechanisation 53, 55–56, 67
  social impacts of mechanisation 53–54, 59–60
  see also Massey University; organic agriculture and horticulture; and under Lincoln College
Howard, Albert 180
Howard, Sir Albert 9, 76
  Agricultural Testament 76
Hudson, Richard 90

Imperial Chemical Industries, Surrey 26
International Federation of Organic Agricultural Movements (IFOAM) 80, 85, 87, 127, 193
  awards night, Theley, 1993 118–20, 121
  conferences 98, 103–04, 106, 109, **110**, 131, 147, 154 (*see also* New Zealand conference, 1994 (below))
  international accreditation programme 107, 111, 117, 120–21, 122, 123, 124, 125, 127, 152
  International Conference on Trade in Organic Products, 1997 154, 156
  New Zealand conference, 1994 126, 129, 130, 131, 134, 135, 136, 138, 141–43, **144**, 145, 148, 149, 152, 153, 173, 179, 197
  New Zealand visit, 1988 114, 118–20, 121, 182
  regionalisation 111, 117, 156, 157
  Register of Internationally Approved Organic Standards 120
  Technical Committee 119–20

World Board 120, 126, 129, 134–35, 141, 149, **151**, 152, 154, 156, 158–59, 188
Intrepid Seeds 101
Ironside, Jeremy **178**
Isle of Wight 15–19, 21
    Whitecliff Bay 24, **27**

Jackson, David 56, 89
Jackson, Trevor 85, 93, 149, 158
Jameson, Elwin **73**
Japanese 'nature' farming 180
Japonica Inn, Springston 75–76
Jarman, Derek 22
Jeffrey, Stuart 138, 140, 159
Jehovah's Witness 16
Jenkins, Tim 85, 131, 179–81, 185–86
Jenkins, Vesna 186
Jill (German shepherd) **78**, 81, **113**, **146**, xi
Joice, Ian 72, **73**
Joliffe, Mr (neighbour, Isle of Wight) 18–19
Jolly, Jim and Eleanor 192
Jones, Margaret 191

Kahl, Holger 104, 157, 158, 161, 172, 186, 188, 189
Kaikōura Peninsula 49
Kain, Bill 161, 167, 177
Kedgley, Sue 188
Kenny, Mr (Bob's teacher) 17
Khen Heng, Shen 163
Kids' Edible Gardens 150, 182
Kirk, Norman 57–58
Kitzsteinhorn, Austria 28, 31
Koukourārata Port Levy 195
Kowhai Farm 168, 170, 180
kūmara industry 44
Kupu, Peter 64–65

La Rooj, Marinus 86, 108
Lake Ida 49
Lambie, Tom 189
Lampkin, Nic 104, 130–31, 141, 145
    Organic Farming 130, 164
    The Landing Restaurant, Christchurch 50, 60, 63, 95
Langley, Leon 50, 60, 63, 67, 81, 95
Lawson, Page 138, 159
le Rooj, Stan 128
Ledebur, Jan von 118, 119, 120
Lester, Fleur **73**
Lilley, Max 130–31

Lincoln College 44, 46–47, 49
    amenity horticulture 75
    Bob's complete retirement 177
    Bob's retirement from lecturing 151–52
    centenary 74
    Department of Horticulture 46, 50–53, 59, 66–67, 72, 74–75, 127
    environmental issues 75, 84
    Horticultural Research Area **48**, 51–52, 66, 80, 88, **89**, 114
    Hudson Hall 49
    Kowhai Farm 168, 170, 180
    negativity towards Bob from some academic staff 10, 100, 122, 128–29, 136–37, 197
    organic horticulture 75, 76, 77, **77**, **78**, 79–80, 83–84, 85
    role in the development of the organic sector 165, 166, 168, 170
    Rose Donaghy shelter belt 128, 138
    'sandwich' degree 75, **77**, 128, **ii**
    seminars on biological husbandry 91–92
    social functions 49–50, 60
    socialist action group 57, 58
    sustainability officer 199
Lincoln College: Biological Husbandry Unit (BHU) 88–89, 93, **99**, **113**, **119**, 121, 122, 128, 137, **139**, 154, **160**, **171**, ii, iii, iv, v, vii, viii, ix, xi, xii, xiii
    biodiversity kits in te reo Māori 196
    Bob's contract to manage 157, 165
    Crowder Block 180
    eco-village vision 157, 158, 196, 198
    expansion 114
    financing 114–15, 131, 148, 149, 164, 166, 168, 195, 196
    food resilience workshop and meeting, 2013 193–94
    lease extension, 2021 196
    Lincoln's clearance of all seed lines and records 170
    Lincoln's use of term Organic Centre 165, 166
    manager 128, 131
    Maples Block 180
    Memorandum of Understanding between university and organic movement, 1999 161–62, 164–65, 166, 167–68, 172, 177
    Michael Watt's interest in supporting BHU 164–68

mission statement 195
OGCT involvement 161
organic community relations 149–50, 174
origins 84, 85, 97–98, 152
publications 148–49
research 96–97, 114, 131, 149
rotation system 114, **132**, 137, **169**, **vii**
soil ability to absorb rainfall **x**
short courses 149
students' views 100–01, 138
technician **102**, 103, 114, 115
Trust management 179–81, 185–86, 189
twenty-first birthday conference 157–58
university review and report, 1995 148–49
unmanaged wilderness after Bob's retirement **175**, **176**, 177
volunteers 9, 100, 138, 140, 148, 157, 158, 172, 174, 181
see also BHU Organic Training College; BHU Organics Trust; Future Farming Centre
Lincoln Environment Organisation 140
Lincoln Tussock Jumpers 143, **144**
Lockeretz, William 9, 87, 88

MacIntyre, Duncan 91
Mallard, Tony 100, 107, 172
Manhire, Jon 101, 109, 145, 161, 162, 179, 186, 188, 194, 195, 196
Maples, Tim 76–77, **78**, 79, 83, 103, 140, 152, 160, 180
Market Development Board 118
Marshall, Denis 143
Martin, Bill 189–90, 195
Maslin, George 80
Mason, Seager 162
Massey University
    Agricultural and Horticultural Systems Management Department 101
    Department of Horticulture 47, 50
May, Chris 86, 87, 88, 90, 91, 97, 104, 105, 107, 108, 113–14, 125, 172, 182–83, 189
May, Jenny 86, 87, 88, 104, 182–83, 189
McArthur, Judith 72–73
McCarthy, J.K., *Patrol into Yesterday* 43
McCarthy show trials 22
McLagan, Rob 84
Measures, Mark 192
Meechin, Angela 75
Meechin, Jack 80, 85–86

Merfield, Charles 128, 140, 157, 158, 162, 163, 166, 181, 195–96, 197
Millton, James 189
Ministry for the Environment 133
Ministry of Agriculture and Fisheries (MAF) 89, 90, 92, 101, 107, 108, 118, 130, 149, 179
    discussion document (1987) 113–14
    MAFQual 117, 125
    MAFTech 112, 117
    organic certification 108, 112, 122, 123–26, 127, 133–34
    Organic Sector Strategy 182
    'Report on biological farming' (1983) 95–96, 112
    research arm 93, 112
    second study of organic farms (1987) 112–13
    support for organic agriculture 111, 112, 122, 128, 133–34
    Sustainable Farming Fund 185
    'Towards sustainable agriculture, organic farming' (1993) 134
Monsanto 52, 55
Montagu, Lord Edward 22
Moore, Mike 106
Moore, Tony 194
Morgan, Tim (later Sol) 129, 138, 184, 187
Morris dancing 12–13, **70**, 71, 72–74, **73**, 81, 100, 121, **144**, 156, 174
Morrison, Mac 46, 50, 67, 75, 98, 129
Moyle, Colin 91, 105–06
Mt Eden, Auckland **32**
Muldoon, Robert 84
Musgrave, David 101, 109, 140, 158, 179

National Association for Sustainable Agriculture Australia (NASAA) 117, 118–19, 121, 125, 157, 182
National Services (Armed Forces) Act 1939 15
Naturkost 120
Nederpelt, Joep 161
New World supermarket, organic food 148
New Zealand Biological Producers and Consumers Council (NZBPCC) 190, 191
New Zealand Biological Producers Council (NZBPC) 104, 105, 111, 112, 113–14, 117, 120–21, 127

establishment and aims 90, 91
and IFOAM international accreditation programme 120–21
MAF's proposed name, Organic Food Standards Board 113
professionalisation 118, 127
relations with the organics community and organisations 106, 107, 108–09, 119
see also BioGro certification programme
New Zealand Fruitgrowers Federation 129–30
New Zealand Journal of Agriculture 46, 52–53, 55, 60, 66, 67
New Zealand Lifestyle Block 187
Newbery, Brian 148
Nightingale, Geoff 26
Nottingham University 25–26
   Ashgate Prize for horticulture 26
   Biological Society trip to France, 1959 26, **27**
   Bob's graduation, 1962 **29, 30,** 31
   The Prodigious Snob performance, c. 1958 **27**
   rowing club 25, 26
   Sutton Bonnington agricultural campus 26, 28
   University Biological Society 26
NZ Organic Suppliers Co. 107–08

Oaro 49
Obervil biodynamic research centre, Switzerland 85
One Tree Hill, Auckland **41**
onion growing and research 45, 51, 52, 55, 56, 66, 72, 136, 138, 181
Opportunity Knocks television show **73**, 74
Organic Advisory Programme (OAP) 188
organic agriculture and horticulture 13, 75–77, 79, 84, 85, 86, 87, 95–96, 122, 142, 168, 199
   global initiative 79, 80–81, 84
   MAF support 133–34
   see also biological husbandry
Organic Christchurch Campaign 150
'organic dilemma' 198
Organic Farm NZ (OFNZ) **183,** 184
Organic Fayre, Christchurch, 1994 142, 143, **144**
Organic Federation of New Zealand (OFoNZ) 170, 182
Organic Garden and Farm Research Center, Pennsylvania 84

Organic Garden City Trust (OGCT) 140, 150, 157–58, 161, 163, 170, 172, 177, 182, 184
Organic Growers Council of New Zealand 88
Organic Growing 90
Organic Matters 168
organic movement 9, 79, 80–81, 84, 87, 91, 103–04, 105, 125, 126, 150, 158, 163, 182, 199
Organic Products and Production Act 2023 199
Organic Products Exporters Group 161, 170
Organic Sector Strategy 182
Organic Training College 188–90, 195, 198–99
Organics 2020, Auckland 182
Organics Aotearoa New Zealand (OANZ) 170, 188, 189, 196
organics education 81, 122, 140, 149, 161, 162, 165, 170, 177, 183, 195, 199
Ōtākaro Orchard 195
Ott, Pierre 135
Otway, Gilda 157
Oxford City Morris 156

Paige, Ralph 135
Palmerston North 47
Panama 35
Papua New Guinea 43
Participatory Guarantee System (PGS) 182–84
Patchett, Brian 109
Patchett, Robyn 172
Patricia Scott School of Ballroom Dancing 25
Pearce, John 90
pelleted seed research 52, 53, 55, 56
Permaculture Association of New Zealand 88
Peryman, Bailey 194–95
pest management 92, 101, 137
pesticides 13, 45, 51, 52, 53, 105, 197
Pesticides Action Network (PAN) 105
Philadelphia 72
Piko Wholefoods 7, 92
Plant Variety Rights legislation 105
Point Chevalier, Auckland **41**
Poland 159
pollution 84, 90, 92, 101, 153
Porteous, Tim 75
Proctor, Peter 108
Project GRO (Giving to Research in Organics) 97, 103, 105, 114, 115, 193, **xii**

psyllid control 195
Public Good Science Fund 150
Pukekohe 45, 46
    youth group **45**, 45–46

Rehu-Murchie, Erihapeti 145
resilient urban systems 193–95
Resource Management Act 1991 133
Richardson, Ruth 106
Robinson, Sir Dove-Meyer 91, 97, 190
Rodale Institute 87
Rodale, J.R. 76
'Rogernomics' 114–15
Ross, Bruce 121, 122
rotation 76, 114, 122, 128, 131, **132**, 137, **169**, 180, **vii**
Round Hill 49
Rowe, Richard 129–30, 131

Salinas Valley, California 53, 63
'sandwich' degree 67, 71, 75, **ii**
Santa Cruz 84
Savory, Allan 136, 145
Schaper, Hans **178**
school gardening programmes 182
Scott, John 86, 90, 92
Scott, Murray 95, **146**, 147, 156
Scott, Robin 108, 111, 112, 122
Seaward Kaikōura Range 49
Seger, Peter 154
Seminar on Biological Farm Management Techniques (Federated Farmers, 1982) 86–87, 88, 89, 90–91, 96
Seyger, Anne 157, 163, 172, 177
Shand, Diana 126
sheep in the rotation 180
Shiva, Vandana 143, 145
Simpson, Anne **xv**
Sims, Bill 72
small farms 12, 54, 135, 181, 182
Small Farms Workshop (BHU Organics Trust) 186
Small-Scale Producers Organic Programme (SSPOP) 182
Smith, Lockwood 150
Snow, Lalange, *War Gardens* (2018) 18
socialism 12, 44, 57–58, 64, 65
Socialist Action 57–58
Soil Association 79, 80, 85–86, 89, 91, 98, 105, 106
    Auckland Branch 86
    Canterbury Branch 76, 80, 92–93, 162
    conferences 90–91, 97
    postgraduate fellowship to Lincoln 90
Soil Association of Great Britain 124
Soil & Health 79, 85–86, 90, 91–92, 93, 95–96, 97, 103, 104, 111, 138
Soil & Health Association 88, 100, 111, 118, 140, 141, 157, 158, 170, 182
    BHU support 115, 149, 161, 162–63, 173
    Bob and Jeanette Fitzsimons as patrons 190–91
    Canterbury Branch 150, 157, 161, 173, 174, 193, 195
    conferences 111–12, 118, 121, 170, 172
Solihull 15, 16, 21
Soltysiak, Ursula 154, 156
Southern Cross 33–36
Soviet Union 55
Spellerberg, Ian 162
Spiller, Perry 96, 98, 105, 108, 109, 113–14, 118, 119, 120, 125, 126, 152, 189
spiritualism 16
Springett, Jo 113, 124
St Catherine's Lighthouse, Isle of Wight 17
Stanhay seed drill 47, 50, 52, 55
Steering Group for Biological Husbandries, 1983 96
Steiner, Rudolf 180
Stephenson, Ian 86, 90
Stevenson, Kim 122, 123, 124, 125
Stevenson screen 22, **xii**
Stokes, Martin 71, 134, **155**, 156, 159, 192
Straight Furrow 86, 87
sublimated sexual desire 22
Sustainable Cities Trust 158
sustainable development 133
Sutton, Jim **183**, 184

Tahiti 36
Tane, Haikai (formerly Ron Mathieson) 75–76, 79, 167, **167**, 170, 179
Tanner, Jon 189
Taranaki, Mt **38**
Tauranga Community College 85
Taylor, Martin 26
te ao Māori worldviews 190
Telford Rural Polytechnic 188
Templer, Michael 86
Tende, France 26

Teshame, Takele 156
Tetu, Brian and Cathy 75
Thiele, Graeme, strawberry production research plots i
Thoma, Klaus 187–88
Thoma, Maria 187, 188
Thompson, Ian 55, 56–57, 64, 65–66
Thompson, Valerie 95–96
tomato growing and research 51, 52, 53, 56, 60, 66, 67, 72, 84
   Bob bottling tomatoes in Carlton Mill Road kitchen **68**
Tonga 64–65
Townsend, Peter 145
Toye, Jonathan 107
Trachta-Dyson, Jamie 164
trade in organic products 111, 112, 118, 130, 133, 143, 152, 154, 196; *see also* exports
Tree Crops Association 193

Unitec Hortecology Sanctuary Mahi Whenua 170, 184, 186
United Nations 'Earth Summit', Rio de Janeiro, 1992 133, 196
United Services Hotel, Christchurch 50
United States Department of Agriculture 81, 84
University of Bath, School of Biological Sciences 67–69, 71, 128
University of California 53–54, 72
   Berkeley 53, 54
   Davis 53, 54
   Small Farm Centre 54
University of Canterbury 49
   Concerned Academics Committee 57
Upton, Simon 86, 106

Vegetable and Potato Growers Federation 130–31
Verity, Rex 162
Victoria Coffee Lounge, Christchurch 63
Vietnam War 44, 58, 59, 65
   protests 13, 54, 57, 59
Vogtmann, Hardy 84–85, 103, 104, 111–12, 135, 145
volunteers 9, 100, 106, 108, 121, 127, 138, 140, 148, 157, 158, 172, 174, 181, 187, 193

Waimakariri District Council 193
Waimauku 44
Waipara tomato trials 56
Waitakere Ranges 40, 44
Walker, Tom 77
Ward, Joe 52
Warren, Miles, flats at 64 Carlton Mill Road, Christchurch 62, 63, 72, 74, 81
Watt, Gaynor 165
Watt, Michael 164–68
Watts, Meriel 182
Waugh, Peter 86–87, 88, 90, 92, 96, 111, 173
weather extremes
   'Big Snow', Christchurch, 1992 131
   Britain, 1947 17–18
weed control 128
Weeks, Chris 71, 74, 81, 147, 192
Wellington 36
Wells, Peter 195
West, Tony 83–84
Whare Flat Folk Festival, Dunedin 74
Whatman, Tony 173
White, Jared 131, 138, **139**, 189, **xi**
White, Lily **178**
Whitelaw, Jack 80
Wilcox, Henry 47, 50
Williams, Morgan 145, 179
Wilson, Hugh 177
Witzenhausen, Germany 84–85
Wood, Frank 165–66, 168, 172, 173
Woods, Dave 86
Woodward, Lawrence 103, 145
Work and Income New Zealand 188
World Trade Organization 154
World War II 15–17, 37, 43
Wratten, Steve 166, 168, 181
Wright, Ray 157, 158
Wwoof Scheme 115, 138, 140, **141**, 159, 162, 187
Wynen, Els 125
Wynn-Williams, Ivo 164–65, 166, 168

Young Anglican and Youth Committee 42, 46
Young, Ken 56
Youngberg, Garth 103–04

Zydenbos, Rick 128